AN EXCEL COMPANION FOR AN INTRODUCTORY STATISTICS COURSE IN SOCIAL AND BEHAVIORAL SCIENCES

First Edition

AN EXCEL COMPANION FOR AN INTRODUCTORY STATISTICS COURSE IN SOCIAL AND BEHAVIORAL SCIENCES

By Alejandro A. Lazarte
Auburn University

Bassim Hamadeh, CEO and Publisher
Tony Paese, Project Editor
Christian Berk, Production Editor
Emely Villavicencio, Senior Graphic Designer
Stephanie Kohl, Licensing Associate
Natalie Piccotti, Director of Marketing
Kassie Graves, Vice President of Editorial
Jamie Giganti, Director of Academic Publishing

Printed in the United States of America.

CONTENTS

PREFACE

Using data to help us present evidence and make decisions is pervasive in science, in professional areas, in public policy, in the media, and in everyday life. Although data can take several forms, numerical values are the most common medium to represent information about observed characteristics of people, events, or objects such as IQ scores, grade point average (GPA) scores, batting averages, number of "likes" in a Facebook page, or percentage of people favoring a particular policy. Disciplines such as mathematics, statistics, computer sciences, decision-making, and psychometrics, contribute with techniques and procedures for dealing with data.

This book is a companion text that introduces you to the use of Excel for performing basic and intermediate data analyses that are common in the social and behavioral sciences. This is not a statistics textbook per se, although the introduction of each chapter contains a summary of the relevant theory behind the Excel computations. This companion focuses on using Excel to perform the most frequent analysis that you can find in introductory textbooks for Statistics for Social and Behavioral Sciences. Thus, it covers computations of descriptive statistics, hypothesis testing for means up to one-way ANOVA, correlation, and simple linear regression. All Excel computations described in this companion rely on basic Excel functions and on the Data Analysis ToolPak© add-on that comes within Excel. In some chapters (mainly when testing means and ANOVA), we complement the analysis by using Excel templates that can be downloaded, together with the Excel files with data and examples for each chapter, from https://cognella.box.com/s/ggkewupycjq36g739mq51mxd3acnb510.

This book presents commands, menus, and shortcuts for Microsoft® Excel® 2016 for Windows®. Excel for Macs® is slightly different in some of the menus and shortcuts. Below we provide some Excel Mac equivalences for the Excel Windows menus and shortcuts used in the book. The Mac equivalences are listed according to the chapter in which the menus and shortcuts appear for the first time.

SOME EXCEL 2016 FOR MAC EQUIVALENCES

Chapter 1

How to highlight cells:

Click on the first cell. Locate the cursor on the square marker at the lower-right corner of the cell and click on the trackpad while swiping to highlight the desired cells.

How to use function keys:

To use function keys, press the function key, *fn*, at the bottom row of the keyboard and the desired function key at the top of the keyboard.

Changing relative to absolute range reference:

Press Command+T

Equivalence for the mouse right-button click:

Clicking the right button of the mouse is equivalent to pressing the *control* key and clicking on the trackpad.

How to open the Name Manager:

To check for all named ranges in the workbook: Press *Command + fn + F3*

Chapter 3

How to install the Data Analysis ToolPak:

To install the Data Analysis ToolPak, click on *Tools* in the Excel top menu. Select *Excel Add-ins.* Check the *Analysis ToolPak* box and click on OK.

Chapter 4

Histogram tool in the Chart Menu:

To change interval width or the number of intervals, click twice on the plot, this opens the "Format Data Series" menu. In the "Bins" box, select "Bin width" or "Number of bins" options.

To add the count values on top of the bars, click twice on the plot. At the "Chart Design" icons, click on "Add Chart Element," go to option "Data Labels" and select "Outside End."

Histogram with Pivot Chart:

By default, Mac Excel will create a histogram after selecting the Pivot Table, we click to insert a Pivot Chart.

To add the count values on top of the bars, click twice on the plot. Click at the "Design" tab next to the "PivotChartAnalyze" tab. Click on "Add Chart Element," go to option "Data Labels" and select "Outside End."

Frequency Polygon with Pivot Chart:

After getting the default histogram chart, click at the "Design" tab next to the "PivotChartAnalyze" tab. Click on "Change Chart type" icon. Select the "Line" option and the first 2-D line alternative.

Cumulative Percentage Polygon with Pivot Chart:

After getting the default histogram chart (the one with cumulative percentages), click at the "Design" tab next to the "PivotChartAnalyze" tab. Click on "Change Chart type" icon. Select the "Line" option and the first 2-D line alternative.

Pie Chart with Pivot Chart:

After getting the default histogram chart, click at the "Design" tab next to the "PivotChartAnalyze" tab. Click on "Change Chart type" icon. Select the "Pie" option and the first 2-D Pie alternative.

Chapter 5

Means and Counts by categories using Pivot Tables:

When creating a table with averages and counts by college year: After dragging GPA to the Values box, to change the entry from Sum to Average, click on the (i) next to the "Sum of GPA" in the "Values" box. In the PivotTable field, select Summarized by Average. Drag again GPA to the Values box and follow a similar procedure to change the Sum into Count.

Chapter 6

Add five-number values to a Box Plot:

Click on the plot. Click on the "Chart Design" tab in the top menu. Click on the "Add Chart Element" icon. Go to the "Data Labels" option and select "Right."

Standard deviations by categories using Pivot Tables:

After dragging GPA to the Values box, to change the entry from Sum to Variance, click on the (i) next to the "Sum of GPA" in the "Values" box. In the PivotTable field, select Summarized by Var. Drag again GPA to the Values box and follow a similar procedure to change the Sum into StdDev.

Chapter 9

Refreshing random values:

Press *fn* + *F9*

Percentages by cell, row, and column using Pivot Tables:

In Excel worksheet 5, to create a single table with percentages per categories: After dragging the "Major" header to the Values box, the default measure is "count." To change to percentage, click on the (i) next to the "Count of Major" in the "Values" box. In the PivotTable field, select "Show data as" and select "% of Column Total."

When creating the cross-tabulations per major and gender: For cell percent, to change the counts to percentage, click on the (i) next to the "Count of Major" in the "Values" box. In the PivotTable field, select "Show data as" and select "% of Grand Total."

Chapters 17 and 18

Scatterplot with simple regression equation:

After getting the scatterplot, click on the "Chart Design" tab in the top menu. Click on the "Quick Layout" and select the first layout.

To add regression line, click on the scatterplot and select the "Chart Design" tab. Click on the "Add Chart Element."

Go to the "Trendline" option and select "More Trendline Options." Check "Linear" and check the "Display Equation on chart" and the "Display R-squared values on chart" boxes.

1 Brief Introduction to Excel

Argumentation with numerical data is a characteristic feature of STEM disciplines and social and behavioral sciences. However, it is common to use quantitative facts in public policy, the media, and in everyday life. Useful numerical data has a meaningful interpretation because they represent observed characteristics of people, events, or objects that are relevant for answering questions. For example:

- Are we getting more cognitively complex in the twenty-first century? Maybe an analysis of the average American College Test (ACT) or Scholastic Aptitude Test (SAT) tests scores of college applicants during the last 50 years can help to answer this question.
- Is the human body reaching its maximum physical performance? Checking the last Olympic times in track and field competitions may provide some evidence for or against.
- Does the number of times we wake up during the night affect our performance in class the next day? Instructors may collect the times students wake up from their Fitbit watches and relate that value to a morning quiz in class.
- Are people that "like" a particular YouTube video all living in the same part of the country? Up to certain point, is it possible to map IP addresses to geographic areas, and simply tally how many likes come from different areas of the country.

Some data is easily accessible from the internet (i.e., the national ACT averages and Olympic records), other data has to be generated using experiments or surveys, and Excel plays a surprisingly important role in all these cases. Due to the widespread availability of Excel in modern times, an Excel file has become a common standard form to store, share, and compute data in sciences, business, and personal applications. An Excel file is a set of ***worksheets*** divided into ***cells***. A cell address consists of a letter that identifies the ***cell column***, and a number, identifying the ***cell row***. In social and behavioral sciences, most of the time, the rows in our data contain data for persons and the columns contain measures for different features for those persons. In addition to containing the data, Excel also has several computational ***functions*** and ***formulas*** that allow us to perform computations, making Excel a useful statistical tool for basic data analysis.

1.1 OBJECTIVES

In this chapter, we start our study of Excel for basic data analysis in social and behavioral sciences by going over the following objectives:

1. To master the basic terminology for addressing contents in an Excel worksheet (cells, ranges) and for using functions and formulas in Excel.
2. To use absolute and relative references for cells when applying functions of formulas across rows or columns.
3. To name cell ranges to facilitate repeated computations.
4. To insert headers, labels, and text in a worksheet.

1.2 BASIC TERMINOLOGY

Excel is a computerized spreadsheet for entering numerical or text data and to perform diverse computations that are continuously actualized when the data changes. Although Excel was designed originally for performing accounting or business record keeping, it has functions and tools with data analysis capabilities.

An Excel file consists of several ***spreadsheets***, labeled by default "Sheet1," "Sheet2," etc. Each sheet is divided into rows labeled with numbers and columns labeled with letters. Thus, a ***cell*** in a spreadsheet has an address in the form of "Letter-Number," i.e., cell C5. A ***range of cells*** is represented as two cells separated by a colon, e.g., A1:A5

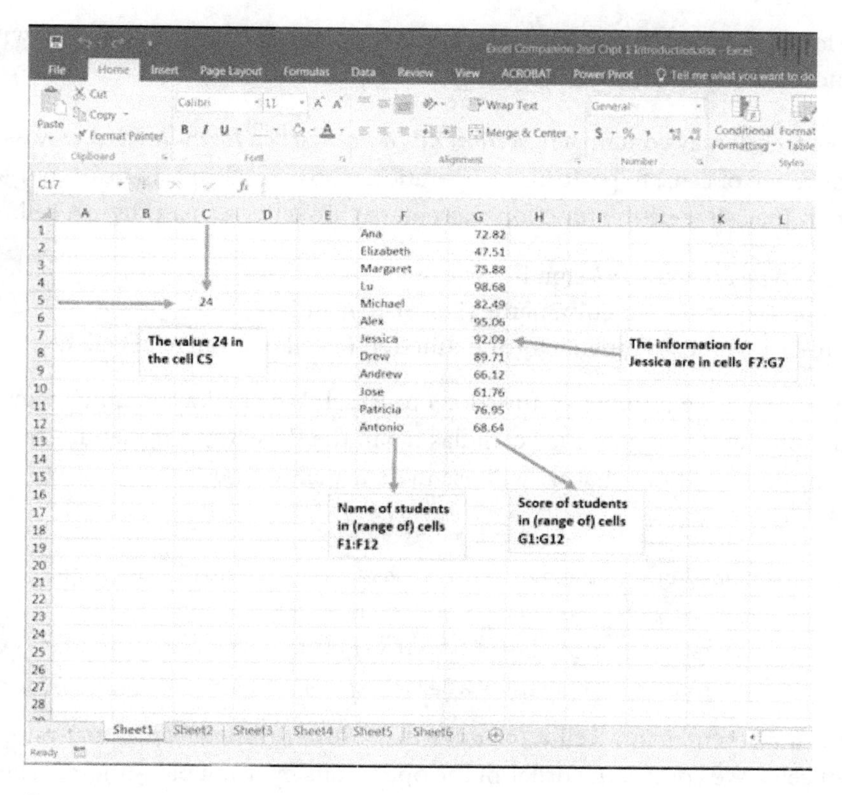

Figure 1.1.

are the first five cells in column A. At the bottom of an Excel worksheet display, there are tabs to facilitate moving among the different spreadsheets of a file.

For example, in the first worksheet of the file EXCEL COMPANION CHPT 1 INTRODUCTION.XLSX you will find (see Figure 1.1):

- the value 24 in cell C5,
- the names of 12 students in the range of cells F1:F12 (i.e., cells from F1 to F12),
- the scores for the 12 students in column G defined by the range G1:G12, and
- the information for Jessica (i.e., her name and her score) are in the range of cells F7:G7.

We can introduce annotations to the worksheet by inserting textboxes:

- Click on the Insert tab in the top menu of the Excel screen and then click on the icon for Text Box.

- Move the cursor to the place where we want to insert the text box, click with the mouse, and then we start entering the desired text.

An Excel cell may contain a ***function, i.e., a predefined formula***, that uses as input a cell, or range of cells, to perform an operation whose resulting single output will be entered in the cell were the function was entered. To enter a function in a cell:

- first, type an equal (=) sign in the cell;
- second, type the function name (i.e., SUM); and
- third, type the arguments for the function (a cell or a range of cells).

There are a large number of functions in Excel. We can have a quick view of all the functions by clicking on the Formulas tab in the Excel top bar menu. Among the most common functions that we will use in data analysis are:

SUM	It adds all the values in a range of cells.
AVERAGE	It computes the mean of a range of cells.
STDEV.S	It computes the sample standard deviation of a range of cells.
COUNT	It returns the number of cells in a range that contain numerical values.
ROUND	It rounds the content of a cell to a specified number of digits or decimals.

We can also enter in a cell a ***formula*** that uses arithmetic operators on cells or range of cells. We control the order of the operations by using parentheses. The common basic operations are:

+	Addition	–	Subtraction	*	Multiplication	/	Division	^	Exponentiation

For example:

In the second worksheet in the Excel file for this chapter, we have the names of 12 students (in column A) and their final scores in a course (in column B). To go to the second worksheet, simply click on the tab "Sheet2" at the bottom of the Excel screen. To put the mean of the students' scores in cell B15 ...

- We can directly type in cell B15 the expression, =AVERAGE(B1:B15) and press enter. The value of the average will appear in the cell (see Figure 1.2).
- It is easier sometimes to type first the equal sign =, then the name of the function, AVERAGE, and then a right parenthesis. At this point, highlight all the cells we want to average (this will insert automatically the range inside the function). Finally, type a right parenthesis and press enter (see Figure 1.3).

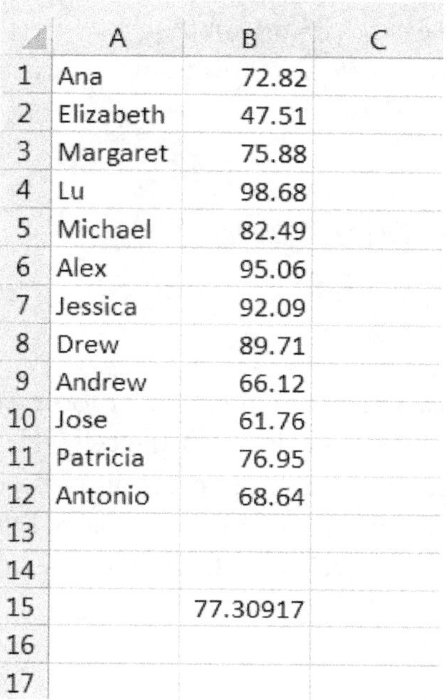

	A	B	C
1	Ana	72.82	
2	Elizabeth	47.51	
3	Margaret	75.88	
4	Lu	98.68	
5	Michael	82.49	
6	Alex	95.06	
7	Jessica	92.09	
8	Drew	89.71	
9	Andrew	66.12	
10	Jose	61.76	
11	Patricia	76.95	
12	Antonio	68.64	
13			
14			
15		77.30917	
16			
17			

Figure 1.2.

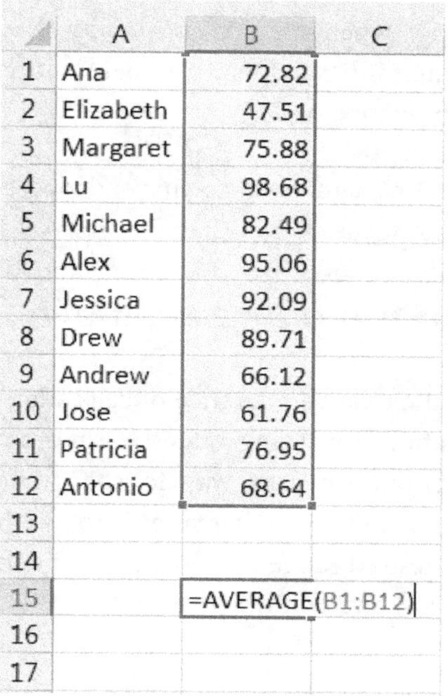

	A	B	C
1	Ana	72.82	
2	Elizabeth	47.51	
3	Margaret	75.88	
4	Lu	98.68	
5	Michael	82.49	
6	Alex	95.06	
7	Jessica	92.09	
8	Drew	89.71	
9	Andrew	66.12	
10	Jose	61.76	
11	Patricia	76.95	
12	Antonio	68.64	
13			
14			
15		=AVERAGE(B1:B12)	
16			
17			

Figure 1.3.

Suppose now that the maximum possible score was 120. To express the average score as a percentage of the possible total, we can enter in cell B16 a formula:

- Type in cell B16 the formula =100*(B15/120) (see Figures 1.4 and 1.5).
- The cell address B15 can be typed directly, or can be inserted by selecting that cell while writing the formula.

Thus, the average score is approximately 64% of the maximum possible score. Notice that when we type in a cell or range of cells in a function, the corresponding cells are also highlighted.

1.3 ABSOLUTE AND RELATIVE CELL REFERENCE

We can refer to a cell or range of cells using an ***absolute*** or ***relative*** cell reference. A ***relative reference*** adjusts to new surroundings when we copy, repeat, or move formulas from one section of the spreadsheet to another. An ***absolute reference*** does not change when copying or moving the cells. We easily identify an absolute reference because it has a dollar sign before its column letter, before its row number, or before both. For example:

Relative reference for column and row, i.e.,	C5
Absolute reference for column and relative reference for row, i.e.,	$C5
Relative reference for column and absolute reference for row, i.e.,	C$5
Absolute reference for column and row, i.e.,	$C$5

The relative reference of cells allows for an easy application of similar formula or function to varying values in rows or columns by simply dragging the cell with the function. An easy way to change from a relative reference to an absolute one (i.e., adding the dollar sign) is to use the key ***F4*** on the keyboard. First type, or select, the desired range of cells and then simply press F4. Excel will add the dollar signs to the row and column locations.

	A	B	C
1	Ana	72.82	
2	Elizabeth	47.51	
3	Margaret	75.88	
4	Lu	98.68	
5	Michael	82.49	
6	Alex	95.06	
7	Jessica	92.09	
8	Drew	89.71	
9	Andrew	66.12	
10	Jose	61.76	
11	Patricia	76.95	
12	Antonio	68.64	
13			
14			
15		77.30917	
16		=100*(B15/120)	
17			

Figure 1.4.

	A	B	C
1	Ana	72.82	
2	Elizabeth	47.51	
3	Margaret	75.88	
4	Lu	98.68	
5	Michael	82.49	
6	Alex	95.06	
7	Jessica	92.09	
8	Drew	89.71	
9	Andrew	66.12	
10	Jose	61.76	
11	Patricia	76.95	
12	Antonio	68.64	
13			
14			
15		77.30917	
16		64.42431	
17			

Figure 1.5.

Example: Relative Reference for Adding a Bonus to Each Student

Suppose we want to add a bonus of 10% to each student's score in column B in the second worksheet. We want to put the new scores in column C. In the third worksheet in the Excel file for this chapter, we have the results.

- For the first student, Ana, we enter in C1 the formula =1.1*B1 and press enter (see Figure 1.6).
- In the lower right corner of cell C1 there is a small square. We drag this square for all students' entries, i.e., until cell C12. The formula being applied to the cells change, i.e., for Ana cell C1 is 1.1*B1; however, for Elizabeth cell C2 is 1.1*B2, for Margaret cell C3 is 1.1*B3, and so on. Thus, the formula correctly uses each student's score to compute her/his additional 10% (see Figures 1.7 and 1.8)

	A	B	C	D
1	Ana	72.82	=1.1*B1	
2	Elizabeth	47.51		
3	Margaret	75.88		
4	Lu	98.68		
5	Michael	82.49		
6	Alex	95.06		
7	Jessica	92.09		
8	Drew	89.71		
9	Andrew	66.12		
10	Jose	61.76		
11	Patricia	76.95		
12	Antonio	68.64		
13				
14				
15		77.30917		
16		64.42431		
17				

Figure 1.6.

	A	B	C	D
1	Ana	72.82	80.102	
2	Elizabeth	47.51		
3	Margaret	75.88		
4	Lu	98.68		
5	Michael	82.49		
6	Alex	95.06		
7	Jessica	92.09		
8	Drew	89.71		
9	Andrew	66.12		
10	Jose	61.76		
11	Patricia	76.95		
12	Antonio	68.64		
13				
14				
15		77.30917		
16		64.42431		
17				

Figure 1.7.

	A	B	C	D
1	Ana	72.82	80.102	
2	Elizabeth	47.51	52.261	
3	Margaret	75.88	83.468	
4	Lu	98.68	108.548	
5	Michael	82.49	90.739	
6	Alex	95.06	104.566	
7	Jessica	92.09	101.299	
8	Drew	89.71	98.681	
9	Andrew	66.12	72.732	
10	Jose	61.76	67.936	
11	Patricia	76.95	84.645	
12	Antonio	68.64	75.504	
13				
14				
15		77.30917		
16		64.42431		
17				

Figure 1.8.

Example: Absolute Reference for Subtracting the Class Average from Each Student

When we want a formula to use the same cell argument for all students, we use absolute references. For example, we want to subtract the class average (that we computed in cell B15) from each of the students' scores.

- For the first student, Ana: we enter in cell D1 an equal sign, then we select or type cell B1, type the minus sign, select cell B15, and finally we press the F4 key to change the relative reference B15 to the absolute reference $B$15. The formula in cell D1 will then be =B1-$B$15 (see Figure 1.9).
- As before, we drag the cell formula to the remaining cells (see Figure 1.10).

The formula changes only one argument, the corresponding to each student's score, but keep the cell with the mean as an absolute argument. Thus, for Ana cell

	A	B	C	D	E
1	Ana	72.82	80.102	=B1-$B$15	
2	Elizabeth	47.51	52.261		
3	Margaret	75.88	83.468		
4	Lu	98.68	108.548		
5	Michael	82.49	90.739		
6	Alex	95.06	104.566		
7	Jessica	92.09	101.299		
8	Drew	89.71	98.681		
9	Andrew	66.12	72.732		
10	Jose	61.76	67.936		
11	Patricia	76.95	84.645		
12	Antonio	68.64	75.504		
13					
14					
15		77.30917			
16		64.42431			
17					

Figure 1.9.

	A	B	C	D	E
1	Ana	72.82	80.102	-4.48917	
2	Elizabeth	47.51	52.261	-29.7992	
3	Margaret	75.88	83.468	-1.42917	
4	Lu	98.68	108.548	21.37083	
5	Michael	82.49	90.739	5.180833	
6	Alex	95.06	104.566	17.75083	
7	Jessica	92.09	101.299	14.78083	
8	Drew	89.71	98.681	12.40083	
9	Andrew	66.12	72.732	-11.1892	
10	Jose	61.76	67.936	-15.5492	
11	Patricia	76.95	84.645	-0.35917	
12	Antonio	68.64	75.504	-8.66917	
13					
14					
15		77.30917			
16		64.42431			
17					

Figure 1.10.

D1 is B1-$B$15, for Elizabeth cell D2 is B2-$B$15, for Margaret cell D3 is B3-$B$15, and so on.

1.4 NAMING A CELL OR RANGE OF CELLS

Giving a name to a range of cells is a convenient way to refer to the same data without selecting the cells continuously. This is a time-saving feature, especially if we will use the same range frequently. Notice that when we give a name to a range of cells, we are implicitly using an absolute reference for the range. The easiest way to name cells is as follows:

- First, highlight the cells or range of cells that you want to name.
- Second, with the cursor on one of the cells, right click the mouse and select the "Define name …" option.
- This will open a little menu. In the "name" box, enter a name for the range of cells. Click OK (see Figure 1.11).

For Example

In the third worksheet in the Excel file for this chapter, we want to name the students' scores in B1:B12 with the name "Scores."

- First, we select the cells.
- Then, with the cursor on the one of the cells, right click the mouse and select the "Define name" option.
- In the New Name window, enter as name "Scores" and then click OK.

Notice that in the "Refers to:" box in the New Name window, the range of values is an absolute reference that includes also the worksheet number.

Excel keeps track of all the named data used in an Excel file. We can see all the named cell ranges by using the Name Manager.

- Click on the FORMULAS tab on the Excel top menu.
- Select the Name Manager.
- A list of all named ranges will appear. You can add, edit, or delete the names using this manager (see Figure 1.12).

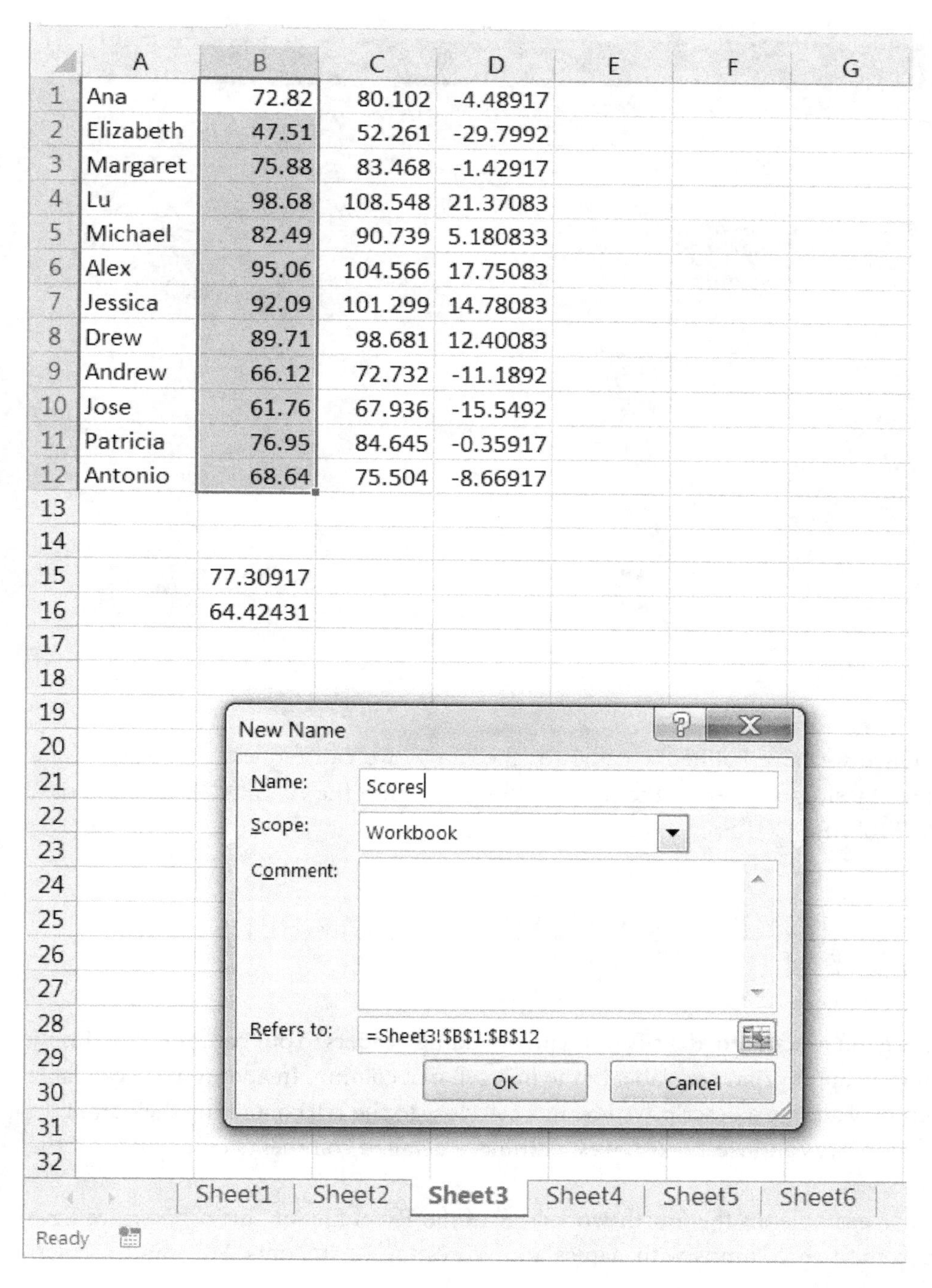
A
B
C
D
E
F
G
1 Ana 72.82 80.102 -4.48917
2 Elizabeth 47.51 52.261 -29.7992
3 Margaret 75.88 83.468 -1.42917
4 Lu 98.68 108.548 21.37083
5 Michael 82.49 90.739 5.180833
6 Alex 95.06 104.566 17.75083
7 Jessica 92.09 101.299 14.78083
8 Drew 89.71 98.681 12.40083
9 Andrew 66.12 72.732 -11.1892
10 Jose 61.76 67.936 -15.5492
11 Patricia 76.95 84.645 -0.35917
12 Antonio 68.64 75.504 -8.66917
15 77.30917
16 64.42431
New Name
Name: Scores
Scope: Workbook
Comment:
Refers to: =Sheet3!$B$1:$B$12
OK
Cancel
Sheet1
Sheet2
Sheet3
Sheet4
Sheet5
Sheet6
Ready

Figure 1.11.

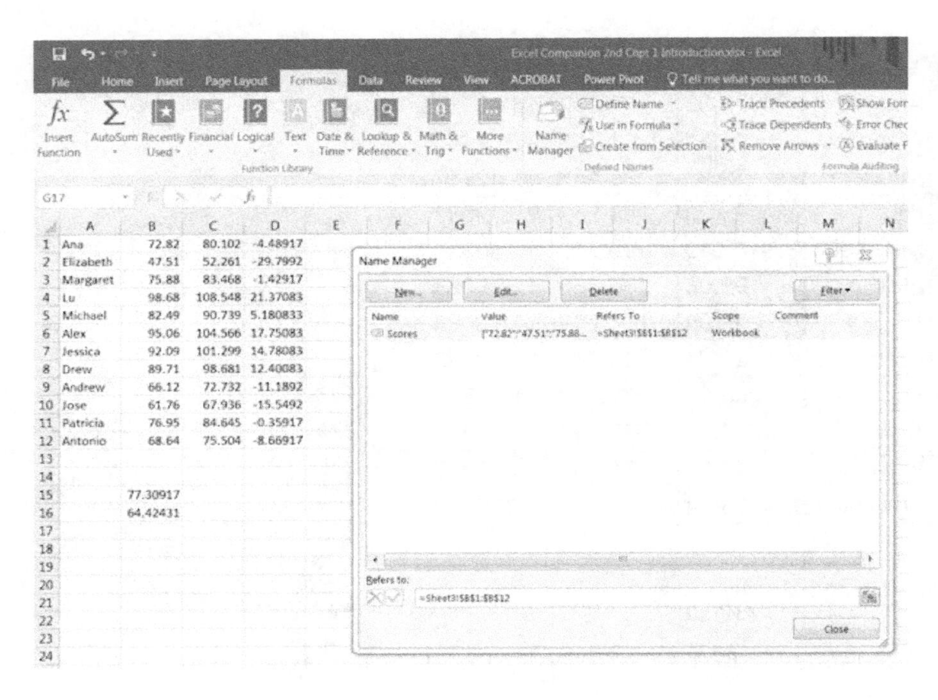

Figure 1.12.

Once we have defined a name for the range, we can request the mean of the scores by simply entering the name of the range into the AVERAGE function, i.e., =AVERAGE(SCORES).

1.5 DESCRIBING RESULTS: HEADERS, LABELS, AND TEXT BOXES

It is a good practice to identify our data by adding headers to our columns. We simply insert an appropriate text label in the first cell of a column. In addition, we can label output by adding a descriptive text in a cell close to the cell containing the output. In addition, we can insert text boxes with more detailed comments on any part of the worksheet.

For example, in the fourth worksheet of the Excel file for this chapter, we have again the two columns with names and scores for 12 students. We want to insert

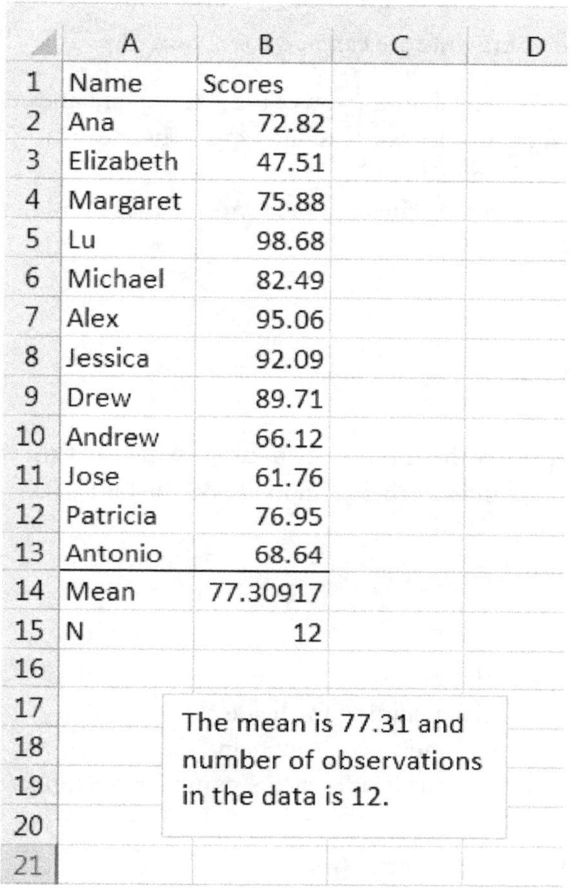

	A	B	C	D
1	Name	Scores		
2	Ana	72.82		
3	Elizabeth	47.51		
4	Margaret	75.88		
5	Lu	98.68		
6	Michael	82.49		
7	Alex	95.06		
8	Jessica	92.09		
9	Drew	89.71		
10	Andrew	66.12		
11	Jose	61.76		
12	Patricia	76.95		
13	Antonio	68.64		
14	Mean	77.30917		
15	N	12		
16				
17				
18				
19				
20				
21				

The mean is 77.31 and number of observations in the data is 12.

Figure 1.13.

headers for the two columns and obtain and label the mean and number of students (see Figure 1.13).

- First, select the first row by clicking on the gray number row 1.
- Right click the mouse and select the "Insert" option to add one more row.
- In the new row, type "Name" and "Score" as headers of the two columns. (You can add border lines below the headers and at the end of the data by using the "Borders" icon in the Font menu.)
- Put the mean score in cell B14, =AVERAGE(B2:B13), and in cell B15 count the number of students, =COUNT(B2:B13).
- To describe these values, we type in A14 "Mean" and in A15 "N," respectively.

To add extensive comments, we can add a text box.

- Click on the INSERT tab in the top of the Excel menu and select "Text Box."
- Drag and open an area for the text and type: "The mean is 77.31 and number of observations in the data is 12."
- We can resize the box by clicking on it and use the round handles on each side of the box.

1.6 WORKED EXAMPLE

Twenty students reported the number of hours per day that they worked in a paid job. Their responses are in worksheet 5 in the Excel file for this chapter (Figure 1.14). The hours are:

5	6	3	3	2	4	7	5	2	3
5	6	5	4	4	3	5	2	5	3

1. Name the range as "Hours":
   - Highlight all the cells, including the header.
   - Right click the mouse and select "Define name."
   - By default, Excel takes the header as the name for the range.

2. Put in cell B23 the sum of the hours:
   - Type in B23, =SUM(Hours) or =SUM(B2:B21).
   - Enter in cell A23 the word **Sum**.

3. Put in cell B24 the number of observations:
   - Type in B24, =COUNT(Hours) or COUNT(B2:B21).
   - Enter in cell A4 the word **Count.**

4. Put in cell B25 the average of the hours:
   - Type the formula in B25, =B23/B24 or =AVERAGE(Hours).
   - Enter in cell A25 the word **Mean.**

	A	B	C	D	E	F	G	H
1		Hours						
2		5	0.9					
3		6	1.9					
4		3	-1.1					
5		3	-1.1					
6		2	-2.1					
7		4	-0.1					
8		7	2.9					
9		5	0.9					
10		2	-2.1					
11		3	-1.1					
12		5	0.9					
13		6	1.9					
14		5	0.9					
15		4	-0.1					
16		4	-0.1					
17		3	-1.1					
18		5	0.9					
19		2	-2.1					
20		5	0.9					
21		3	-1.1					
22								
23	Sum	82						
24	Count	20						
25	Mean	4.1	0.000					
26								
27								

John A. Smith

The average number of hours was 4.1

When the value of the average is subtracted from each observation, the average of the resulting differences is 0

Figure 1.14.

5. In column C, insert the difference between each individual value of hours and the average of the data:
   - For the first subject, type in cell C2, =B2-$B$25. (Immediately after typing B25, press the F4 key. This will insert the dollar signs.)
   - Drag the cell (using the little box in the lower right corner of the cell) to the rest of the rows with data for hours.

6. In cell C25, enter the mean of the differences:
   - In cell C25, type the function =AVERAGE(C2:C21). (You can enter the range by highlighting the values.)
   - The result may appear in scientific notation (3.55271E-16) to change it into a three decimals number.
   - Click on the cell with the mouse right button and select Format Cell.
   - Click the Number tab, select the option "Number" and select three decimal places. Ok.

7. Finally, insert a text box on the right-hand side of your calculations:
   - Click on the INSERT tab in the top of the Excel menu and select "Text Box."
   - Drag and open an area for the text and type "Your name. The average number of hours was ... ... (put the value you got)." "When the average of the data is subtracted from each observation, the average of the resulting differences is ... ... (put the value you got)."

Your final output, as shown in the figure above, is in worksheet 6 in the Excel file for this chapter.

2 Summation in Excel

Although data analysis is a complex endeavor, at a very basic level it consists of a set of computations to summarize the values observed in a group of individuals into fewer values describing the group. Summarization of data nearly always involves, in one way or another, the sum of the original values or a transformation of those values, i.e., a ***summation*** of values. For many analyses, we can list in plain English the computations required to perform the data analysis task and use Excel to perform those computations. However, verbal description of computations is prone to errors, and the most frequent way to present data analysis computations in textbooks, manuals, publications, etc. is by using mathematical expressions. Thus, we will practice how to read and translate summation expressions into Excel computations. For example, a well-known statistic is the arithmetic average. We can state how to compute an average using a summation expression, a plain description, or as an Excel function. For example, suppose that we have 10 values of a variable labeled X, then:

- $\frac{1}{10}\sum_{i=1}^{10} X_i$ is the summation expression.
- "Add all the ten values of X and then divide the total by 10," is the plain English translation.
- =1/10*SUM(A1:A10) will be the Excel translation (the 10 values in cells A1 to A10). Of course, another possibility is =AVERAGE(A1:A10).

Thus, we will treat equations using summation expressions as a convenient way to describe computational instructions that we can implement using Excel functions or commands.

2.1 OBJECTIVES

In this chapter, we will introduce the summation notation and how to translate that notation into computational commands to be implemented using Excel.

1. To identify the components of a summation expression.
2. To translate into English the computational steps expressed in summation notation.
3. To obtain in Excel the computations required by a summation expression.

2.2 SUMMATION NOTATION

A summation notation formula has three basic components:

1. The summation symbol sigma, Σ, that indicates that we have to add the expression that follows.
2. An expression that we have to add, usually with a subscripted index.
3. A range for the subscript, usually with the starting value at the bottom of the sigma and the ending value at the top of the sigma symbol. For example:

$$\sum_{i=1}^{N} X_i$$

The expression that we add is X_i, where X is the name of a variable with several values and the subscript i indicates the specific value in that variable, i.e., X_2 will be the second value for the variable X. The range for the subscript goes from 1, i.e., the first value in X, to N, i.e., the Nth value in X. Traditionally, N represents the total number of values for the variable.

For example, the first worksheet in the file EXCEL COMPANION CHPT 2 SUMMATION.XLSX contains $N = 5$ values for the variables X and Y (column A contains the subscript for each value) (see Figure 2.1). Thus, the summation formula,

	A	B	C
1	i	X	Y
2	1	1	4
3	2	3	3
4	3	1	1
5	4	0	1
6	5	2	0
7		7	9
8			

Figure 2.1.

the corresponding English translation, the corresponding Excel formula, and the result of the computation for adding all the values of each variable are given below:

$\sum_{i=1}^{5} X_i$ "Add all the values (from 1st to 5th) of variable *X*." In cell B7, enter =SUM(B2:B6).

$\sum_{i=1}^{5} X_i = 7$

$\sum_{i=1}^{5} Y_i$ "Add all the values (from 1st to 5th) of variable *Y*." In cell C7, enter =SUM(C2:C6)

$\sum_{i=1}^{5} Y_i = 9$

Of course, there are more complex summation expressions. For example, in the second worksheet in the Excel file for this chapter, we have $N = 5$ values for two variables: *X* and *Y*, and use them to perform the computations below (see Figure 2.2). Again, we present the summation expression, the English translation, and the steps to follow for computing the result in Excel.

$\sum_{i=1}^{5} X_i^2$ "Square each of the five values of *X*. Then, add all the five values."

- We create a new column in D, labeled "X2."
- Enter in cell D2, the equation =B2^2.
- Drag down the formula (using the little box in the lower right of cell D2) to fill the other cells up to D6.
- Finally, we enter the sum function in cell D7, i.e., SUM(D2:D6).

	A	B	C	D	E	F
1	i	X	Y	X2	4Y+2	XY
2	1	1	4	1	18	4
3	2	3	3	9	14	9
4	3	1	1	1	6	1
5	4	0	1	0	6	0
6	5	2	0	4	2	0
7		7	9	15	46	14
8						

Figure 2.2.

- A better way to obtain the sum of the squared values of X is to use the function =SUMSQ( ). With this function, we do not need to create another column with the squared values of X in order to obtain the sum of-squared values. Thus, =SUMSQ(B2:B6) will produce the same total 15 that we found before.

The result is $\sum_{i=1}^{5} X_i^2 = 15$

$\sum_{i=1}^{5}(4Y_i + 2)$ "Multiply each value of Y by 4 and add 2 to the result. Finally, add all the five values."

- We create a new column in E with label "4Y+2."
- We enter in cell E2, the expression =4*C2+2.
- Drag the expression to fill other cells up to E6.
- We enter the sum of the values in Cell E7, i.e., =SUM(E2:E7).

The result is $\sum_{i=1}^{5}(4Y_i + 2) = 46$

$\sum_{i=1}^{5} X_i Y_i$ "Multiply the corresponding values of X and Y in each row. Then, add all the five resulting values."

- We enter in column F as header "XY."
- In cell F2, we enter =B2*C2.
- We drag the formula in F2 to the remaining rows.
- Finally, we enter in cell F7 the sum of the values, i.e., SUM(F2:F7).

The result is $\sum_{i=1}^{5} X_i Y_i = 14$

RAW, DEVIATION, AND STANDARD SCORES

2.3

Two of the most common transformations of *raw scores* or the original individual scores in data analysis are *deviation scores*, when we subtract the mean from each score, and *standard scores*, when we subtract the mean from each score and divide the result by the standard deviation (SD). We will talk in more detail about SD in a later chapter; now we will focus on using that value in a summation computation.

Deviation score: We subtract the data average (represented by X-bar) from each of the individual observations.

$$D_i = X_i - \overline{X}$$

Standard scores: We subtract the data average from each of the observations and divide the result by the SD of the data (represented by the letter S).

$$Z_i = \frac{X_i - \overline{X}}{S}$$

In the third worksheet in the Excel file for this chapter, we have five raw scores for the variable *X* in column B (see Figure 2.3). We will add two new columns with the deviation and standard scores. First, we find the mean and the SD for the raw scores *X*.

	A	B	C	D
1	i	X	D Score	Z Score
2	1	1	-0.4	-0.3508232
3	2	3	1.6	1.40329283
4	3	1	-0.4	-0.3508232
5	4	0	-1.4	-1.2278812
6	5	2	0.6	0.52623481
7	Mean	1.4		
8	SD	1.140175		
9	Formula 1		0	0
10	Formula 2		1.140175	1
11				

Figure 2.3.

- In cell B7, enter the function =AVERAGE(B2:B6).
- In cell B8, enter the function =STDEV.S(B2:B6).

In column C, we obtain the deviation scores:

- In cell C1, enter as label "D Score."
- In cell C2, we subtract the mean of X from the first value of X, i.e., =B2-$B$7. Notice that we use an absolute cell reference for B7 (use the F4 key to insert the dollar signs).
- Drag the function in C2 to the remaining rows.

In column D, we insert the standard scores:

- In cell D1, enter as label "Z Score."
- In cell D2, we subtract the mean of X from the first value of X and then we divide the difference by the SD, i.e., = (B2-$B$7)/$B$8. Again, we use an absolute reference for the cells with the mean and the SD.
- Drag the function in D2 to the remaining rows.

Other examples of summation formulas include finding the mean and SD (by simply squaring the values) for the new transformed scores. For example:

Formula 1: $\frac{1}{N}\sum_{i=1}^{N} D_i \qquad \frac{1}{N}\sum_{i=1}^{N} Z_i$

In English: "Add all the values and then multiply the total by one over the number of values (in our case $N = 5$)."

- For D scores: In cell C9, enter the following equation: = 1/5*SUM(C2:C6) or AVERAGE(C2:C6).
- For Z scores: In cell D9, enter the following equation: = 1/5*SUM(D2:D6) or AVERAGE(D2:D6).

Formula 2: $\sqrt{\frac{\sum_{i=1}^{N} D_i^2}{N-1}} \qquad \sqrt{\frac{\sum_{i=1}^{N} Z_i^2}{N-1}}$

In words: "Square each value and add all of them. Then, divide the total by the number of values minus one. Finally, take the square root of the result."

- For D scores: In cell C10, enter the following equation: = SQRT(SUMSQ (C2:C6)/4).
- For Z scores: In cell D10, enter the following equation: = SQRT(SUMSQ (D2:D6)/4).

Formula 1 computes the average of the *D* and *Z* scores, which always will be zero for deviation and standard scores. Formula 2 computes in this case (where the means are zero) the SD for the *D* and *Z* scores. For D scores, Equation 2 produces the same value as the SD of the raw scores *X*, but for standard scores, the SD will always be 1. We will talk about these properties of standard scores in a later chapter.

MORE THAN ONE WAY TO OBTAIN THE SAME SUM 2.4

A source of confusion when dealing with equations in data analysis is that sometimes there is more than one-way to compute the same value, and, therefore, more than one-way to represent the same result in summation notation. For example, the two summation expressions below produce the same result.

$$\text{(a)} \quad \sum_{i=1}^{N}(X_i - \bar{X})^2 \qquad \text{and} \qquad \text{(b)} \quad \sum_{i=1}^{n}X_i^2 - \left(\sum_{i=1}^{n}X_i\right)^2 \Big/ N$$

However, the steps for the computation are different. Expression (a) requests to subtract the average from each one of the observations in *X* and square the difference before we add all the results. Expression (b), on the other hand, requests to obtain the sum of *X*, and the sum of X^2 first. Afterward, we subtract from the sum of X^2 the square of the sum of *X* divided by the number of values *N*.

In the fourth worksheet in the Excel file for this chapter, we perform these operations (see Figure 2.4). In cells B2 to B6, we have again the five raw scores for *X*.

- In cell B7, we compute the sum of the raw scores *X* by entering the function =SUM(B2:B6).
- In cell B8, we obtain the sum of the squared values of *X* by using the function =SUMSQ(B2:B6).
- In cell B9, we compute the average of *X* using the function =AVERAGE(B2:B6).
- In cell B10, we get the number of values *N* using the function =COUNT(B2:B6).
- In column C, we enter the squared values of the difference scores. In cell D2, we enter the function =(B2-$B$9)^2, and drag the function to the rest of the rows.
- Finally, in cell C7, we put the sum of the values, i.e., =SUM(C2:C6).
- Thus, the value 5.2 has been computed using expression (a).
- To compute expression (b), we put in cell C12, the expression = B8-(B7^2)/B10, and we obtain the same value 5.2. Notice that we do not actually need to type the cells. After entering the equal sign, we click on the cell containing the sum

	A	B	C	D
1	i	X	(X-Xbar)2	
2	1	1	0.16	
3	2	3	2.56	
4	3	1	0.16	
5	4	0	1.96	
6	5	2	0.36	
7	Sum	7	5.2	(a)
8	SumSq	15		
9	Mean	1.4		
10	N	5		
11				
12	(b)	5.2		
13				

Figure 2.4.

of squares (B8) and its name is inserted into the equation we are writing. In a similar way, we click on the other cells when we want to refer to them in our equation. The equation in C12 represents the following summation equation:

$$\sum_{i=1}^{n} X_i^2 - \left(\sum_{i=1}^{n} X_i\right)^2 \Big/ N = 15 - (7)^2/5 = 5.2$$

2.5 WORKED EXAMPLE

The number of hours (variable X) that 20 students slept the night before an examination is below.

8	8	7	7	8	6	8	8	9	9
5.5	6	9	7	8	5	7	7	9	9

	A	B	C	D
1	X	(X-Xbar)2		
2	8	0.225625	N	20
3	8	0.225625	Sum X	150.5
4	7	0.275625	Sum X2	1161.25
5	7	0.275625	Mean	7.525
6	8	0.225625	SSx	28.7375
7	6	2.325625	Var	1.596528
8	8	0.225625	SD	1.263538
9	8	0.225625		
10	9	2.175625		
11	9	2.175625		
12	5.5	4.100625		
13	6	2.325625		
14	9	2.175625		
15	7	0.275625		
16	8	0.225625		
17	5	6.375625		
18	7	0.275625		
19	7	0.275625		
20	9	2.175625		
21	9	2.175625		
22				

Figure 2.5.

These values will be in column A, from A2 to A21, in the fifth worksheet of the Excel file for this chapter (see Figure 2.5).

1. Find the number of values in the data in column A, i.e., *N*
    - In cell D2, enter =COUNT(A2:A21).
    - In cell C2, enter the label "N."

2. Find the sum of X, or $\sum_{i=1}^{N} X_i$
    - In cell D3, enter =SUM(A2:A21).
    - In cell C3, enter as label "Sum X."
3. Find the sum of X squared, or $\sum_{i=1}^{N} X_i^2$
    - In cell D4, enter =SUMSQ(B2:B21).
    - In cell C4, enter as label "Sum X2."
4. Find the mean of X, or $\overline{X}$
    - In cell D5, enter =AVERAGE(A2:A21).
    - In cell C5, enter as label "Mean."
5. Find the following expression called "sum of squares": $SS_X = \sum_{i=1}^{N} (X_i - \overline{X})^2$
    - In cell B2, we enter the equation for the value in A2, subtract the mean, and square the result, i.e., we enter =(A2-$D$5)^2. (Notice that D5 contains the mean, and it is an absolute reference.)
    - Drag the formula on the remaining cells. Enter "(X-Xbar)2" as header in C1.
    - In cell D6, enter =SUM(C2: C21), and in cell C6, enter as label "SS*x*."
6. Find the following expression called "variance," $S_X^2 = \frac{SS_X}{N-1}$
    - In cell D7, enter =D6/(D2-1).
    - Enter in C7 as label "Var."
7. Find the following expression called "SD," $S_X = \sqrt{S_X^2}$
    - In cell D8, enter =SQRT(D6).
    - Enter in cell C8, enter as label "SD."

The final spreadsheet, after all the computations, appears in the sixth worksheet of the Excel file for this chapter.

3 Frequency Distributions, Data Analysis ToolPak, and Pivot Tables

It is increasingly common to find expressions such as *Data Analysis, Big Data, Data Mining,* etc. Data in those expressions may refer to information in a wide range of formats, i.e., numerical, text, sound, visual, etc. Thus, data are the ACT scores for all female students applying to college in 2019, a list of the references mentioned in a scientific article, files containing 5 seconds of bird songs, or jpg picture files with brain images of a patient. We will focus in this manual on ***numerical data***, i.e., information that is codified using numbers, and in this chapter on how to describe the occurrence of those numerical values. We will see that a ***frequency distribution***, **the tally of the frequency in which range of values occur in the data**, will enable us to answer questions such as:

- Is the country getting older? What are the most common ages of people living in the United States in 2019?
- Is Professor Garcia's students' ACT score distribution similar to Professor Smith's students' scores?
- Of all the lunches that I had last year, how many of them cost me over 10 dollars?

Excel has two ways to provide answers to questions such as the ones above. One is using the ***Data Analysis ToolPak***, that is an Excel Add-in for performing statistical analyses, and the other is using Excel ***Pivot Tables***.

3.1 OBJECTIVES

In this chapter, we will introduce the Data Analysis ToolPak and Pivot Tables in the context of learning how to obtain frequency distributions.

1. To identify the components of a table of frequency distributions.
2. To distinguish between frequency, cumulative frequency, percent, and cumulative percent tables.
3. To install Data Analysis ToolPak in Excel and to use it for frequency distributions computation.
4. To apply pivot tables for obtaining frequency distributions.

3.2 FREQUENCY DISTRIBUTIONS AND THE DATA ANALYSIS TOOLPAK

A ***frequency distribution*** is **a table that provides the frequency of occurrence of individual values or intervals of values in a data set**. A frequency distribution table consists of a set of ***intervals*** or ***bins*** in which the total ***range***, or **difference between the largest and smallest value** in the data has been partitioned. The number of intervals is a function of the desired ***interval width*** or **the range of values for each bin**. Each interval is represented by a ***lower and upper limit***, i.e., 10–30. We usually start listing the intervals by the lower limit of the lowest interval (i.e., the interval for the lowest values in the distribution).

Although Excel is not a software specifically aimed for statistical analysis, it provides an Add-in with templates to perform several of the most common statistical data analysis tasks. The ***Data Analysis ToolPak***, although always included among the Excel program files, is not installed by default on the Excel menu. If your Excel version has indeed the Data Analysis ToolPak installed, you will see a **"Data Analysis" icon** when you select the **Data tab** in the top Excel menu, as illustrated below (Figure 3.1).

If there is not a "Data Analysis" icon, you need to install the Add-in according to the following:

- In the Excel top menu click "File" and then select "Options" on the left-hand side bar.
- In the Excel options menu click on "Adds-ins." At the bottom of the available Adds-ins there is a "Manage" box in which you have to select "Excel Add-ins" and then click the "Go" button.

- A new Add-ins menu will appear with the available "Add-ins" (see Figure 3.2); check the boxes for the "Analysis ToolPak" and, optionally, the "Analysis ToolPak-VBA." Finally click OK.

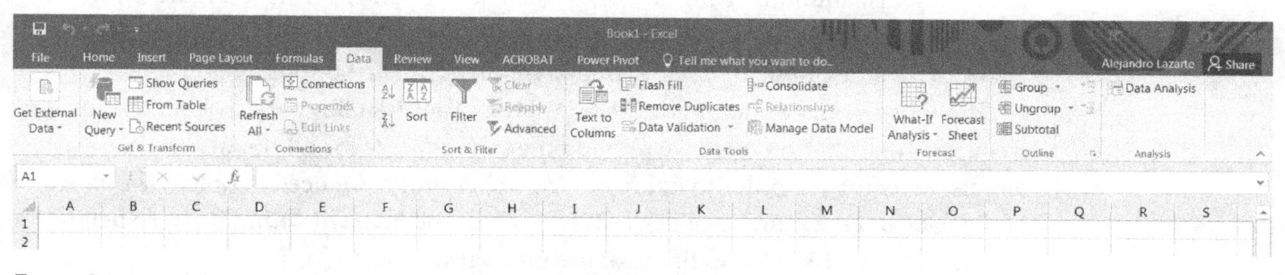

Figure 3.1.

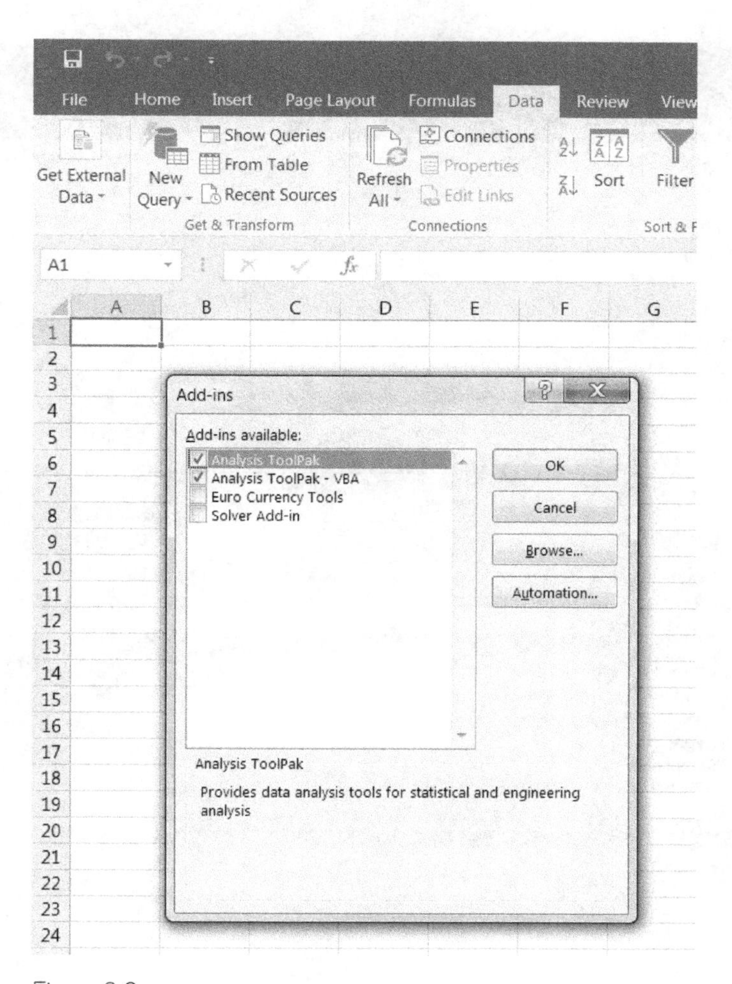

Figure 3.2.

Once we have the "Data Analysis" icon, we can click on it in order to have access to a menu with a list of statistical procedures. In the present example, we will obtain frequency distributions using the option labeled "Histogram" (see Figure 3.3). We will use the data in the file EXCEL COMPANION CHPT 3 FREQ DISTRIBUTION. XLSX to illustrate how to obtain grouped frequency distributions.

The "Histogram" option in the Data Analysis ToolPak allows us to obtain grouped frequency distribution once we provide the ***upper limits*** for all the desired class intervals. Evidently, in order to figure out the upper limits, we need to obtain the ***range*** of the data and the number of desired intervals or the ***interval width***. It is also helpful to complete the list of lower limits of the intervals.

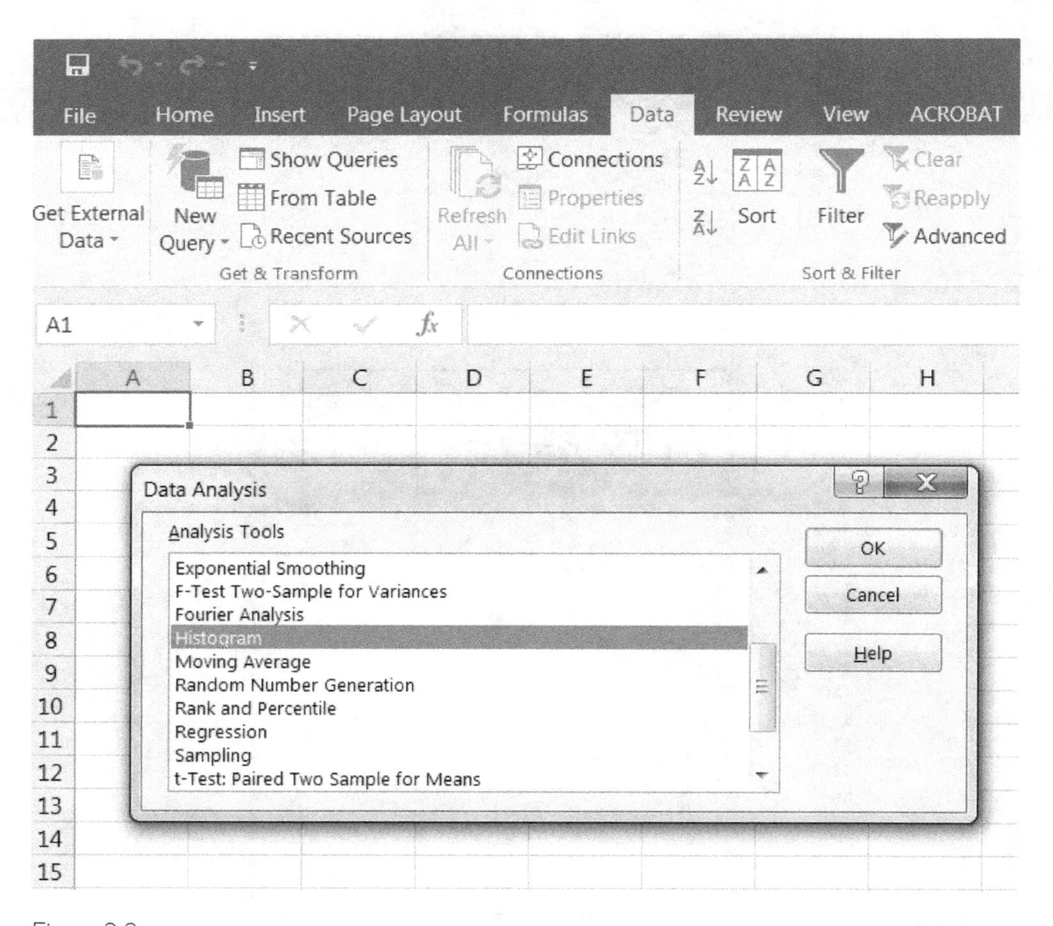

Figure 3.3.

	A	B	C	D	E	F	G	H
1	**Scores**					**Lower Limit**	**Upper Limit**	
2	80		**Min=**	35		30	39	
3	73		**Max=**	94		40	49	
4	51		**Range=**	59		50	59	
5	81		**N=**	35		60	69	
6	46		Nintevals=	6		70	79	
7	85		Apro Width	9.833333		80	89	
8	84		Width*	10		90	99	
9	75							
10	44							
11	84							

Figure 3.4.

In column A of the first worksheet of the Excel file for this chapter, we have 35 students' final scores in a course in the range A2:A36 (see Figure 3.4). The total number of students is in cell D5 (we find this number using =COUNT(A2:A36) and we want to get a frequency distribution for these values.

- We need to find the range of the values. First, we put the data minimum value in D2 by using the function =MIN(A2:A36) and the maximum value in D3 with the function =MAX(A2:A36). In our case, these values are 35 and 94, respectively.
- Using those values, we obtain the spread or range of the values in D4 by subtracting the minimum from the maximum, =D3 – D2. In this case, this range is 59 points.

We have to choose an appropriate number of intervals. For example, for six intervals, we find the approximate interval width:

- Type in D6 the desired number intervals, i.e., 6 in our case.
- Then, the approximate interval width will be the range of the data divided by the desired number of intervals. In cell D7, we enter =D4/D6. In our case it is 9.8333.

- Because number of intervals should be an integer value, we roundup the value in D7 using the function ROUNDUP. In D8, we enter =ROUNDUP(D4, 0), where 0 means that we want no decimals.
- Notice that we can try different values for the Nintervals in D6 and observe how the interval width changes.

For the lowest interval in the table, we can use as lower limit for that interval any sensible value. Suppose that we decide to use 30 as the lower limit for the lowest interval.

- Enter 30 in cell F2.
- The lower limit for the next interval will be 40 (remember we are using an interval width of 10) thus, we enter 40 in the next cell F3.
- To fill the remaining lower limit values, we highlight the two cells with 30 and 40 and drag them to the remaining five rows in that column (until getting the value of 90). This will create the sequence 30, 40, 50, …, etc.
- Repeat the same procedure for the upper limits. Enter 39 in cell G2 and 49 in the next cell G3. Highlight both and drag them to filling the other values of upper limit.

Now, we are ready to use the Data Analysis ToolPak.

- Click on the DATA tab at the top Excel menu and then click on the Data Analysis icon.
- In the Data Analysis window select "Histogram" and click OK.
- In the Histogram menu (see Figure 3.5), enter the data range in the "Input Range" box, in our case A2:A36.
- Enter the upper limits in the "Bin Range" box, in our case G2:G8.
- In the output options, we can create the table in a new worksheet, or in the same worksheet by selecting "Output Range." In our example, we request the table in the same worksheet, with the upper-left corner of the table in cell D13.
- Finally, we check the Cumulative Percentage box, and press OK.

In worksheet 2 of the Excel file for this chapter, we have the frequency distribution table created by the Histogram menu (see Figure 3.6). The table has a column with the bin values, which are actually the upper limits of our intervals, a column with the frequencies, and a column with the cumulative percentages. Thus, there are nine

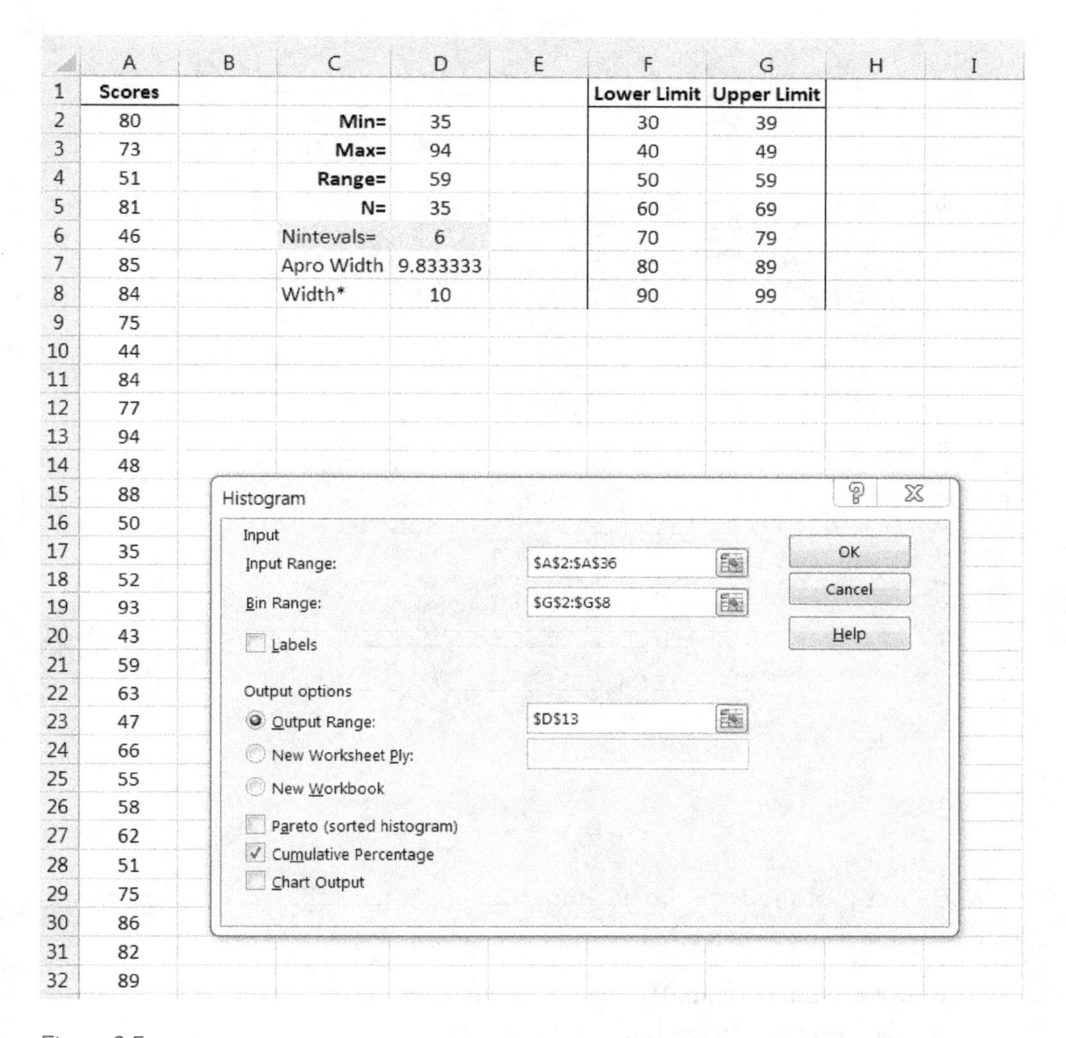

Figure 3.5.

students in the interval 50–59 and also nine students in the interval 80–89. Also, 51.34% of the students have scores of 69 or less.

The frequency distribution table still requires some editing. We can modify the table by editing the class intervals in the Bin column, and by adding two additional columns: One expressing the occurrence of the values in each interval as a percentage of the 35 students, and the other with the cumulative frequencies for the intervals.

	A	B	C	D	E	F	G	H	I
1	**Scores**					**Lower Limit**	**Upper Limit**		
2	80		**Min=**	35		30	39		
3	73		**Max=**	94		40	49		
4	51		**Range=**	59		50	59		
5	81		**N=**	35		60	69		
6	46		Nintevals=	6		70	79		
7	85		Apro Width	9.8333333		80	89		
8	84		Width*	10		90	99		
9	75								
10	44								
11	84								
12	77								
13	94			*Bin*	*Frequency*	*Cumulative %*			
14	48			39	1	2.86%			
15	88			49	5	17.14%			
16	50			59	9	42.86%			
17	35			69	3	51.43%			
18	52			79	6	68.57%			
19	93			89	9	94.29%			
20	43			99	2	100.00%			
21	59			More	0	100.00%			
22	63								
23	47								
24	66								

Figure 3.6.

Worksheet 3 of the Excel file for this chapter contains the modified frequency distribution table (see Figure 3.7).

- We edit the Bin column. We type as new header "Interval," and in each row we type the lower and upper limit for the interval.
- For the cumulative frequency column: copy in cell G14 the first frequency of the table (i.e., the contents of cell E14).
- In cell G15, enter the equation =G14+E15 (which provides the sum of the first two frequencies, 6 in our case).
- Finally, drag this equation to the remaining rows. We can add a "Cum.Freq" header to the column.
- For the percentages, enter in H14 the ratio of the frequency of the first frequency cell over 35 (the total sample size), i.e., =E14/35.
- Then, drag this equation to the remaining cells in the column.

	A	B	C	D	E	F	G	H	I
1	**Scores**					**Lower Limit**	**Upper Limit**		
2	80		**Min=**	35		30	39		
3	73		**Max=**	94		40	49		
4	51		**Range=**	59		50	59		
5	81		**Count=**	35		60	69		
6	46		Nintevals=	6		70	79		
7	85		Apro Width	9.8333333		80	89		
8	84		Width*	10		90	99		
9	75								
10	44								
11	84								
12	77								
13	94			*Interval*	*Frequency*	*Cumulative %*	*Cum.Freq*	*Percent*	
14	48			30-39	1	2.86%	1	2.86%	
15	88			40-49	5	17.14%	6	14.29%	
16	50			50-59	9	42.86%	15	25.71%	
17	35			60-69	3	51.43%	18	8.57%	
18	52			70-79	6	68.57%	24	17.14%	
19	93			80-89	9	94.29%	33	25.71%	
20	43			90-99	2	100.00%	35	5.71%	
21	59								
22	63								
23	47								
24	66								

Figure 3.7.

- To change the ratios into percentages, highlight all the values in the column, right click on the mouse and select "Format Cells." In the "Number" tag of the menu select "Percentage" and then OK. You can add a "Percent" header to the column.
- Finally, we can delete the last row of the table (with label "More").

FREQUENCY DISTRIBUTIONS USING PIVOT TABLES 3.3

A more efficient way to obtain a frequency distribution in Excel is by using a pivot table. A ***Pivot Table* is an interactive way to create tables with quantitative information (sum, counts, means, frequencies, etc.)**. It is the easier way to obtain ungrouped (i.e., the values are not categorized into class intervals) as well as grouped frequency distributions for numerical values and even for categorical data (i.e., data that can be text or nominal).

In worksheet 4 of the Excel file for this chapter, we have the original score data for the 35 students (with header "Scores"). To create an ***ungrouped frequency distribution***, i.e., **a table that lists each individual value in the data**, we do the following:

- Click on the INSERT tab at the top Excel menu, and then click on the Pivot Table icon.
- This will open a "Create Pivot Table" menu (see Figure 3.8). Select as input the range with the data, **including the header**; thus the range is A1:A36. On the other hand, if you have named the range of values, type in the name in the Table/Range box.

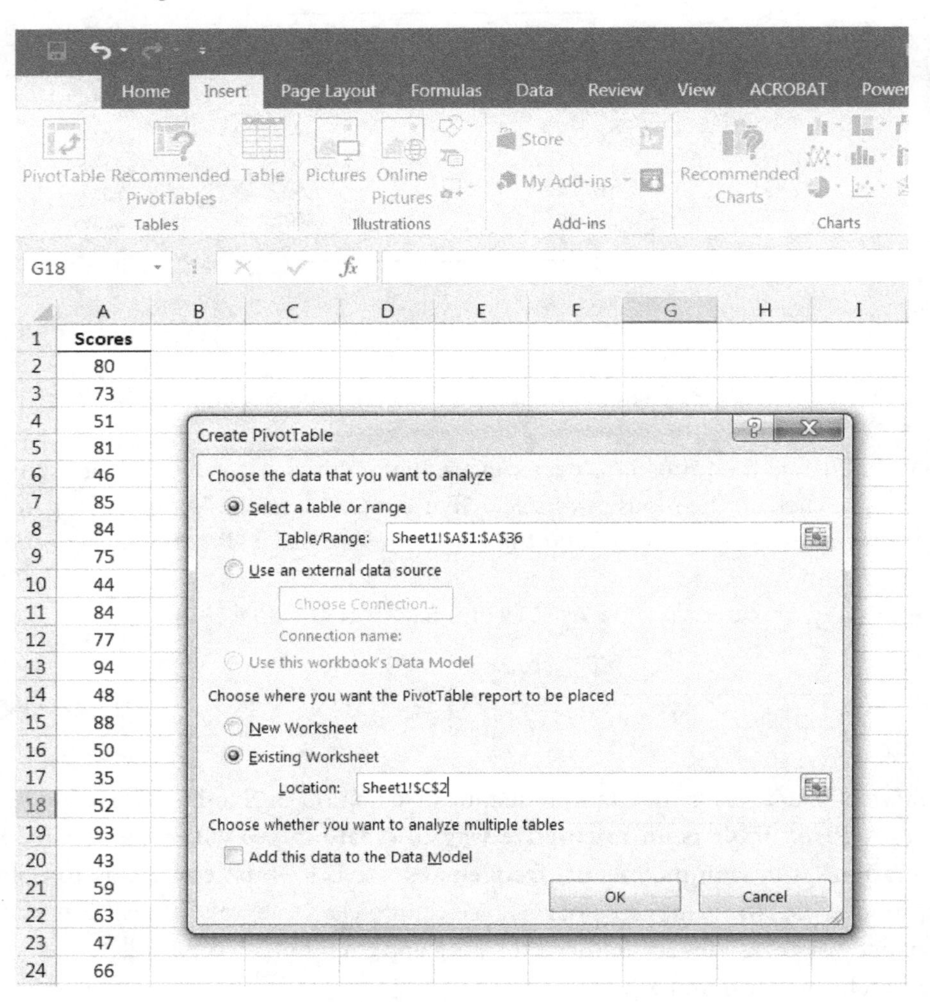

Figure 3.8.

- The pivot table can be located on a new worksheet or in the current worksheet. In our example, click "Existing Worksheet" and enter the location in cell C2; in this way the upper-left corner of the table will be on that cell. Click OK.
- A "Pivot Table Fields" menu will appear on the right-hand side of the screen, as well as an outline for the table in the selected location (see Figure 3.9). The label "Scores" will appear in the top box of the "Pivot Table Fields" menu. Drag that "Scores" label to the "ROWS" box at the bottom of the menu. You will see that Excel creates the first column of the table with "Row Labels" as header.
- Drag again the "Scores" label from the top of the menu to the "VALUES" box at the bottom of the Pivot window. Excel creates the second column, which for numerical values contains by default the sum of the values within each of the rows.
- We need to change the function from sum to count. Right-click on any value of the second column and select "Value Field Settings ..." option in the opening menu. In that menu, select the tab "Summarize values by" and, in the "Summarize value field by," select the option "Count." Finally, click OK.

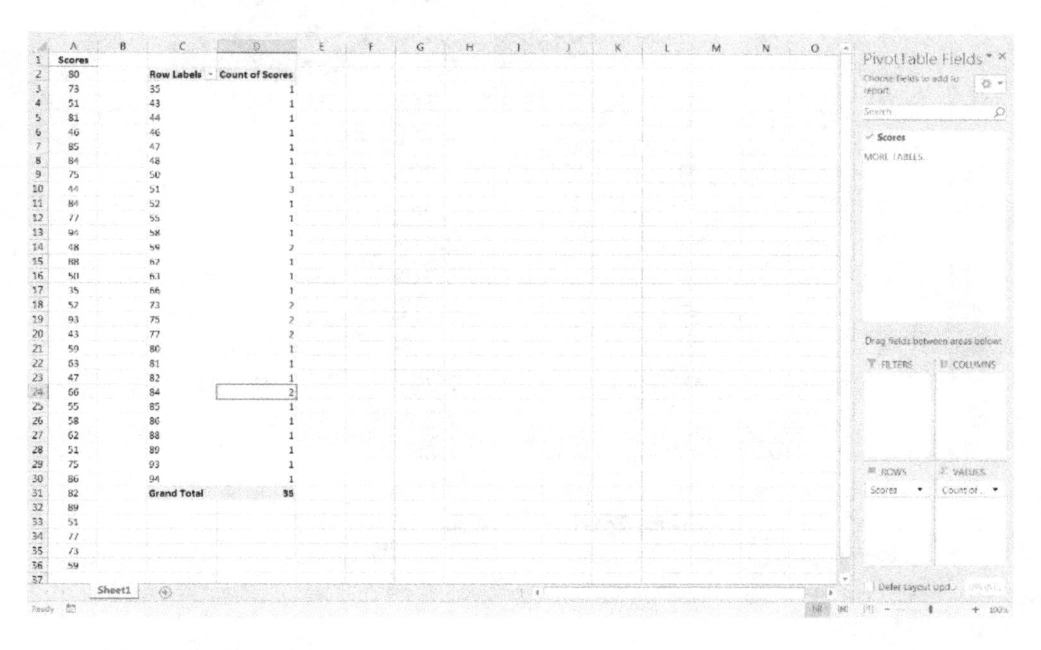

Figure 3.9.

	A	B	C	D	E	F	G	H
1	**Scores**							
2	80		**Scores**	**Freq.**	**Percent**	**Cum.F**	**Cum.%**	
3	73		35	1	2.86%	1	2.86%	
4	51		43	1	2.86%	2	5.71%	
5	81		44	1	2.86%	3	8.57%	
6	46		46	1	2.86%	4	11.43%	
7	85		47	1	2.86%	5	14.29%	
8	84		48	1	2.86%	6	17.14%	
9	75		50	1	2.86%	7	20.00%	
10	44		51	3	8.57%	10	28.57%	
11	84		52	1	2.86%	11	31.43%	
12	77		55	1	2.86%	12	34.29%	
13	94		58	1	2.86%	13	37.14%	
14	48		59	2	5.71%	15	42.86%	
15	88		62	1	2.86%	16	45.71%	
16	50		63	1	2.86%	17	48.57%	
17	35		66	1	2.86%	18	51.43%	
18	52		73	2	5.71%	20	57.14%	
19	93		75	2	5.71%	22	62.86%	
20	43		77	2	5.71%	24	68.57%	
21	59		80	1	2.86%	25	71.43%	
22	63		81	1	2.86%	26	74.29%	
23	47		82	1	2.86%	27	77.14%	
24	66		84	2	5.71%	29	82.86%	
25	55		85	1	2.86%	30	85.71%	
26	58		86	1	2.86%	31	88.57%	
27	62		88	1	2.86%	32	91.43%	
28	51		89	1	2.86%	33	94.29%	
29	75		93	1	2.86%	34	97.14%	
30	86		94	1	2.86%	35	100.00%	
31	82		**Grand Total**	**35**	**100.00%**			
32	89							
33	51							
34	77							
35	73							
36	59							

Figure 3.10.

- To add a column with the relative frequency or percentage of observations in each value, we drag again "Scores" to the VALUES box in the Pivot Table menu. To change the sum to count, we open again the Valued Field Settings menu, but in addition to changing the summarize value to count, click on the tab "Show Value as." In the "Show value as" box, search for the option "% of Column Total." Click OK.
- To add a column with the cumulative frequency at each value, we drag again "Scores" to the VALUES box in the Pivot Table menu. We open the Valued Field Settings menu, change the summarize value to count, and in the "Show Value as" menu, we search for the option "Running Total In." Click OK.
- To add a column with the cumulative percentage at each value, we drag again "Scores" to the VALUES box in the Pivot Table menu. We open the Valued Field Settings menu, change the summarize value to count, and in the "Show Value as" menu, we search for the option "% Running Total In." Click OK.
- The final table has as column labels "Row Labels," "Count of Scores," "Count of Scores2," etc. We can edit these headers by clicking on those headers and replacing the custom name with "Scores," "Freq," "Percent," "Cum.F," and "Cum.%" (see Figure 3.10).

Pivot tables are a more convenient way to obtain ungrouped frequency distribution tables. However, how do we convert this table into a ***grouped frequency distribution***, i.e., **a table with class intervals defined by the user**? This process is quite simple using pivot tables.

- Right click on any of the values in the first column of the table (the Scores column). In the menu that appears, select the option "Group." A grouping menu will appear requesting the starting (lower limit of the lowest interval) and ending values of the distribution (see Figure 3.11). By default this values are the minimum and maximum values in the data. Also, Excel shows a suggested interval width in the box "by:".
- We enter 30 as the starting value and leave the ending value at 94 and the by at 10. Click OK.

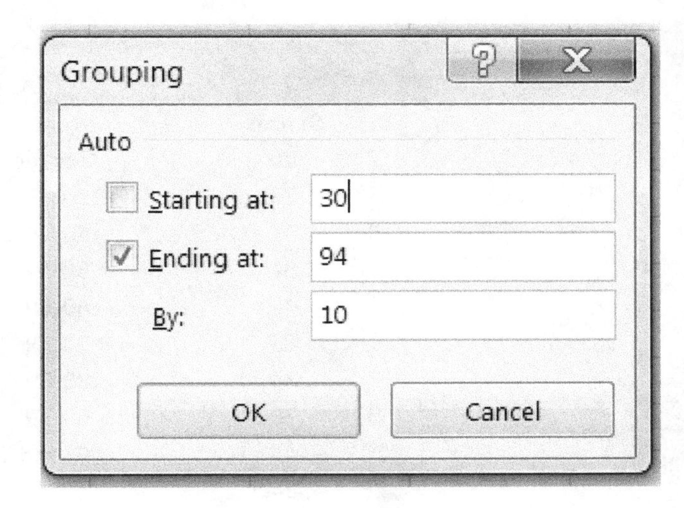

Figure 3.11.

The table will show correctly the intervals, frequency, and percentage, but the cumulative frequency and percentage columns will revert to show frequencies (see Figure 3.12). This happens because the "Show value as" property for the cumulative columns will be lost. We only need to click on those columns and select again the correct "Show value as" property (see Figure 3.13).

	A	B	C	D	E	F	G	H
1	Scores							
2	80		Scores	Freq.	Percent	Cum.F	Cum.%	
3	73		30-39	1	2.86%	1	1	
4	51		40-49	5	14.29%	5	5	
5	81		50-59	9	25.71%	9	9	
6	46		60-69	3	8.57%	3	3	
7	85		70-79	6	17.14%	6	6	
8	84		80-89	9	25.71%	9	9	
9	75		90-99	2	5.71%	2	2	
10	44		Grand Total	35	100.00%	35	35	
11	84							

Figure 3.12.

	A	B	C	D	E	F	G	H
1	Scores							
2	80		Scores	Freq.	Percent	Cum.F	Cum.%	
3	73		30-39	1	2.86%	1	2.86%	
4	51		40-49	5	14.29%	6	17.14%	
5	81		50-59	9	25.71%	15	42.86%	
6	46		60-69	3	8.57%	18	51.43%	
7	85		70-79	6	17.14%	24	68.57%	
8	84		80-89	9	25.71%	33	94.29%	
9	75		90-99	2	5.71%	35	100.00%	
10	44		Grand Total	35	100.00%			
11	84							

Figure 3.13.

One of the advantages of pivot tables is that we can ungroup the table to its original ungrouped form. Again, the frequencies and percentages will be correct, but we will need to redefine the "cumulative" columns by changing again their "Show value" property.

PIVOT TABLES FOR CATEGORICAL DATA

3.4

Pivot tables can easily produce frequency tables for data that is not numeric. For example, in worksheet 5 in the Excel file for this chapter, we have the class membership (Freshman, Junior, etc.) for 74 undergraduate students. Notice that the header is "Year in College." We want to obtain a frequency distribution table for years in college (see Figure 3.14).

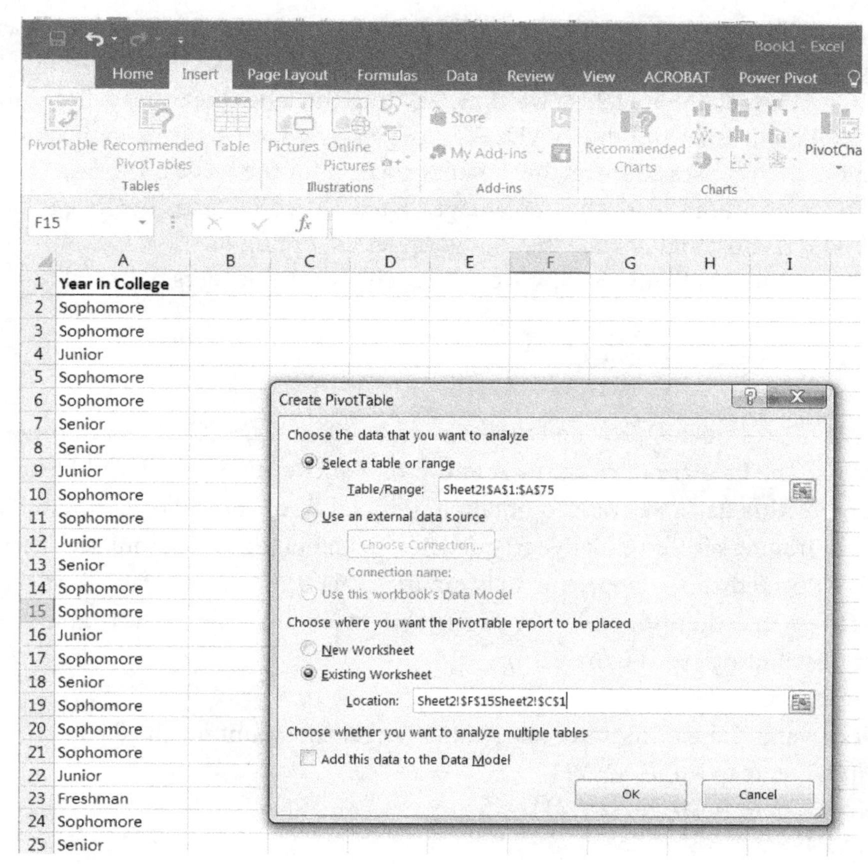

Figure 3.14.

	A	B	C	D	E	F
1	**Year in College**		**Row Labels**	**Count of Year in College**	**Count of Year in College2**	
2	Sophomore		Freshman	2	2.70%	
3	Sophomore		Junior	26	35.14%	
4	Junior		Senior	8	10.81%	
5	Sophomore		Sophomore	38	51.35%	
6	Sophomore		**Grand Total**	**74**	**100.00%**	
7	Senior					
8	Senior					

Figure 3.15.

- Click on the INSERT tab at the top Excel menu. Click on Pivot Table.
- Select as input the range with the data (including the header); in this case A1:A75.
- Click "Existing Worksheet" as the location for the Pivot Table. Enter C1 as the location for the upper-left right corner of the table. Click OK.
- In the "Pivot Table Fields" window, we drag the label "Year in College" to the "ROWS" box and to the "VALUES" box. Notice that when the data is not numeric, Excel uses count as the default function for the values.
- To add a column with the percentages, we drag again "Years in College" to the VALUES box, and change the "Show values as" option to "% of Column Total."
- The result is in Figure 3.15. We can later change the headers to the table at will.

3.5 WORKED EXAMPLE

A sample of 20 students provided their Grade Point Average (GPA) and gender to a researcher. The data are in columns A and B of worksheet 6 in the Excel file for this chapter. For this data, we want to obtain a grouped frequency distribution table for GPA with frequencies, cumulative frequencies, percentages, and cumulative percentages. We decided to use approximately six intervals. We use the Data ToolPak and Pivot Tables. In addition, we use a pivot table to obtain a frequency table for the gender distribution (see Figure 3.16).

1. First, we find the minimum, maximum, range, and count for the data.
   - In E2, enter =MIN(A2:A21).
   - In E3, enter =MAX(A2:A21).
   - In E4, enter =E3–E2.
   - In E6, enter COUNT(A2:A21).

	A	B	C	D	E	F	G	H	I	J	K	L	M	N
1	GPA	Sex					Lower	Upper		Row Labels	Count of GPA	Count of GPA2	Count of GPA3	Count of GPA4
2	2.8	Female		Min=	2		2	2.2		2-2.3	2	10.00%	2	10.00%
3	3.3	Female		Max=	4		2.3	2.5		2.3-2.6	1	5.00%	3	15.00%
4	3	Female		Range=	2		2.6	2.8		2.6-2.9	1	5.00%	4	20.00%
5	4	Male		N=	20		2.9	3.1		2.9-3.2	8	40.00%	12	60.00%
6	2	Male		N Intervals	6		3.2	3.4		3.2-3.5	1	5.00%	13	65.00%
7	3	Female		Apro. Width	0.33333333		3.5	3.7		3.5-3.8	3	15.00%	16	80.00%
8	3.1	Male		Width*	0.3		3.8	4		3.8-4.1	4	20.00%	20	100.00%
9	3.6	Female								Grand Total	20	100.00%		
10	3.8	Female		*Bin*	*Frequency*	*Cumulative %*	*Cum Freq*	*Perce.*						
11	3	Female		2.2	1	5.00%	1	5.00%						
12	3	Female		2.5	2	15.00%	3	10.00%						
13	4	Male		2.8	1	20.00%	4	5.00%						
14	4	Female		3.1	8	60.00%	12	40.00%						
15	2.3	Male		3.4	1	65.00%	13	5.00%						
16	2.9	Female		3.7	3	80.00%	16	15.00%						
17	3.5	Female		4	4	100.00%	20	20.00%						
18	2.4	Male		More	0	100.00%								
19	3	Male												
20	3.1	Female		Row Labels	Count of Sex									
21	3.6	Female		Female	13									
22				Male	7									
23				Grand Total	20									
24														

Figure 3.16.

2. For six intervals what is the approximate bin width and the width rounded up to one decimal?
   - We put 6 in E6 and compute the approximate width in E7 as =E4/E6.
   - We round the approximate width to one decimal. In E8, we compute =ROUND(D7,1). The approximate width is 0.3333. We will use a width of 0.3.
3. Write down the lower and upper limits for the class intervals; start with the minimum value of the data. How many intervals do we actually need?
   - From G2 to G8, we input the lower limits starting at the minimum (2.0) and following 2.3, 2.6, etc.
   - In H2 to H8, we enter the upper limits, starting at 2.2, then 2.5, 2.8, etc.
   - Although we wanted six intervals, we actually needed seven to go up to the maximum value of 4.
4. Use the Data Analysis ToolPak to obtain the grouped frequency distribution. Add a column for cumulative frequency and for percentage. What is the interval with the largest frequency? What is the percentage of students with GPA of 2.8 or less?
   - Using the Histogram option in the Data Analysis tool, enter as Input Range A2:A21, as Bin Range H2:H8, and as output range in cell D10 for the upper left corner of the table.
   - Add in G the column Cum.Freq. In cell G11, enter =E11, in cell G12, enter =E12+G11, and drag this formula to the remaining cells.

- Add in H the column of Perce. In cell H11 enter =E11/$E$5 and drag this formula to the remaining cells. Highlight all the values in the column, right click the mouse and select "Format Cells." In the "Number" tag of the menu select "Percentage" and then OK.
- The most frequent class interval is the 2.9 to 3.1 Twenty percent of students have a GPA of 2.8 or less.

5. Using pivot tables, we obtain the grouped frequency distribution.
   - Click on the Data tab at the top Excel menu and then click on Pivot Table icon.
   - Select as input the range for GPA, A1:A21. Click "Existing Worksheet" as the location for the Pivot Table, and use J1 as the location for the upper-left corner of the table. Click OK.
   - In the "Pivot Table Fields" window, we drag "GPA" to the "ROWS" box and to the "VALUES" box. We change the function to counts. Drag again GPA to the VALUES box and change the function to count and show values to % Column Total.
   - Right click on any value in the first column and select the Group option. Use as starting value 2, as ending value 4, and as a "by" value 0.3. You will notice that the pivot table uses a slightly different set of upper limits from the Histogram option in the Data Analysis ToolPak; however, the counts are very similar.
   - After grouping the values, add the cumulative frequency and the cumulative percentage columns.

6. Use a pivot table to obtain the frequencies for gender. How many students are female?
   - Click on the Data tab at the top Excel menu and then click on Pivot Table icon.
   - Select as input the range B1:B21. Click "Existing Worksheet" as the location for the Pivot Table. Enter D20 as the location for upper-left corner of the table. Click OK.
   - In the "Pivot Table Fields" menu, drag the label "Gender" to the "ROWS" box and to the "VALUES" box. The default function for the values is counts. There are 13 females in the class.

The final output of the analysis is in worksheet 7 of the Excel file for this chapter.

4 Distribution Graphs

As the saying goes, "A picture is worth a thousand words." The conclusions of many data analyses can be quickly and effectively communicated to the users by using graphical representations. In this chapter, we will go over the graphical ways to display the tables of frequency distributions we introduced in the previous chapter. The basic plot is the ***histogram*** that uses bars to represent the frequency or percentages of the class intervals in a frequency distribution table. A ***frequency polygon*** fulfills the same role as histograms, but it uses lines instead of bars to represent the frequency or percentages. A frequency polygon is the preferred option when plotting cumulative percentages. Finally, a ***pie chart*** is a popular plot for displaying the frequency of categorical data using slices of a circle.

There is wide use of these types of graphs:

- How can we quickly visualize the distribution of the scores in a final examination in a class to judge if it is too easy or too hard?
- Can we use a chart to locate the height and weight development of children in comparison to children of the same sex and age?
- How can we represent in a single graph the distribution of age by males and females in a population?
- How do we graph the consumers' preference for a particular type of fast food?

4.1 OBJECTIVES

We will use the Data Analysis ToolPak, Excel charts, and Pivot Charts to pursue the following objectives:

1. Data Analysis ToolPak to create histograms and cumulative frequency polygons.
2. Histogram chart to quickly create histograms.
3. Pivot Tables and Pivot Charts to produce histograms, frequency polygons, and pie charts.

4.2 HISTOGRAMS

Histograms use bars to display the frequency or the percentage of values in the class intervals of a frequency distribution table. Starting with Excel 2016, we have three ways to produce a histogram: (1) using the Data Analysis ToolPak, (2) using the Histogram option from the Charts menu, and (3) using a Pivot Chart. In the Excel spreadsheet, EXCEL COMPANION CHPT 4 DISTRIBUTION GRAPHS.XLSX, we have examples and data for this chapter.

Histogram Using the Data Analysis ToolPak

To obtain a histogram using the Data Analysis ToolPak, we follow the same steps as the ones we used in Chapter 3 for producing a frequency distribution table. In worksheet 1 of the Excel file for this chapter, we have the scores for 35 students and the lower and upper limits for the class intervals we want to use.

- Click on the Data tab in the Excel top menu. On the right-hand side, click on the Data Analysis icon.
- In the Data Analysis window select "Histogram" and click OK.
- In the Histogram menu, enter as input range A2:A36; enter in the Bin range the upper limits in D2:D8, and as Output Range cell G1 (for the upper-left corner of the table). ***In addition***, click the "Chart Output" box option, and then click OK. (see Figure 4.1)

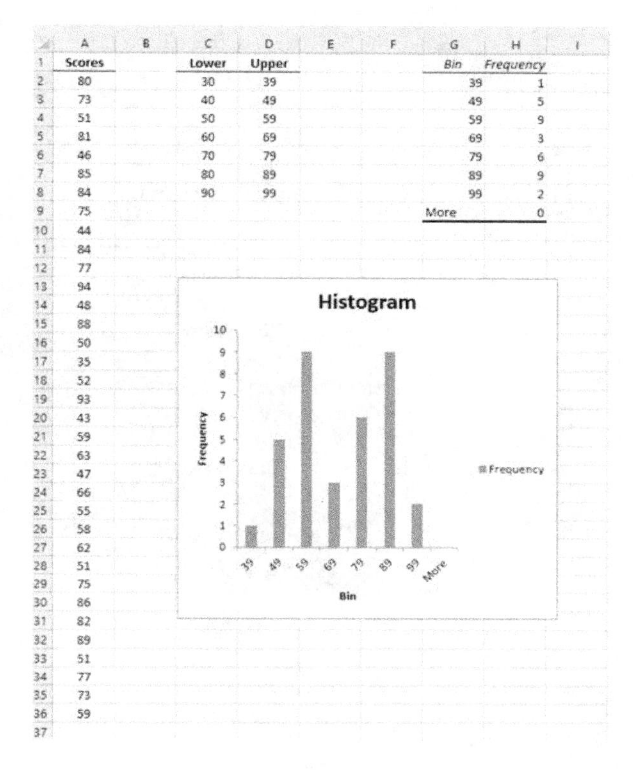

	A	B	C	D	E	F	G	H
1	Scores		Lower	Upper			Bin	Frequency
2	80		30	39			39	1
3	73		40	49			49	5
4	51		50	59			59	9
5	81		60	69			69	3
6	46		70	79			79	6
7	85		80	89			89	9
8	84		90	99			99	2
9	75						More	0
10	44							
11	84							
12	77							
13	94							
14	48							
15	88							
16	50							
17	35							
18	52							
19	93							
20	43							
21	59							
22	63							
23	47							
24	66							
25	55							
26	58							
27	62							
28	51							
29	75							
30	86							
31	82							
32	89							
33	51							
34	77							
35	73							
36	59							
37								

Figure 4.1.

Excel will produce a histogram with wide spaces between the bars, an additional "More" interval, the label "Frequency" for the vertical axis, and "Bin" as the label on the horizontal axis. In addition, there is a caption on the right with the word "Frequency." We need to edit these default options. In worksheet 2 of the Excel file for this chapter, we have the result of the following modifications (Figure 4.2):

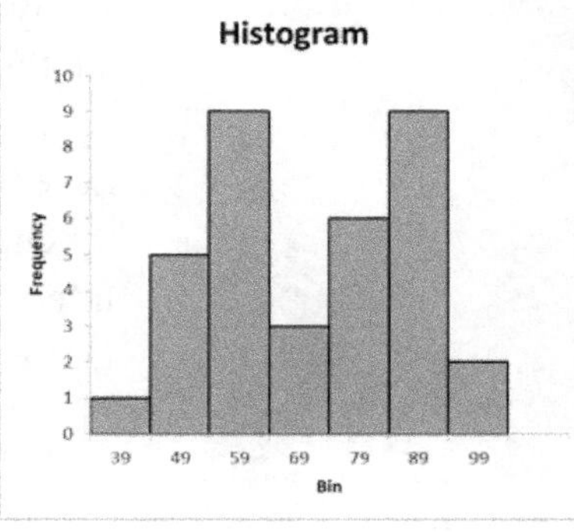

Figure 4.2.

- First, we delete the "Frequency" caption on the right of the plot by selecting and deleting it.
- Second, we eliminate the gaps between the bars (a histogram only has gaps if there are no values in a class interval). Click twice on the bars to open the "Format Data Series" menu. In the Gap Width section, slide the marker to 0% gap. To add borderlines to distinguish the bars, select the Fill and Border options in the same menu (the paint can). In the Border menu, select solid line and black as color, then press close.
- Finally, you can delete the "More" entry in the frequency distribution table.

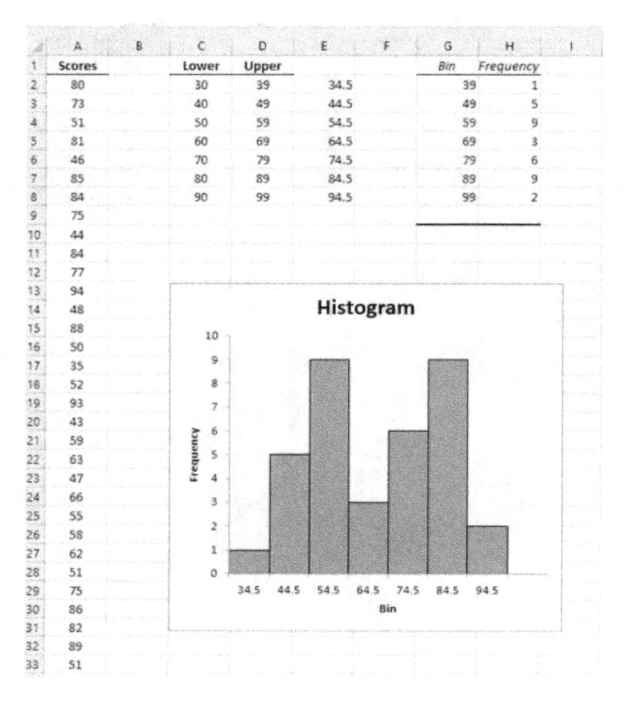

	A	B	C	D	E	F	G	H
1	Scores		Lower	Upper			*Bin*	*Frequency*
2	80		30	39	34.5		39	1
3	73		40	49	44.5		49	5
4	51		50	59	54.5		59	9
5	81		60	69	64.5		69	3
6	46		70	79	74.5		79	6
7	85		80	89	84.5		89	9
8	84		90	99	94.5		99	2
9	75							
10	44							
11	84							
12	77							
13	94							
14	48							
15	88							
16	50							
17	35							
18	52							
19	93							
20	43							
21	59							
22	63							
23	47							
24	66							
25	55							
26	58							
27	62							
28	51							
29	75							
30	86							
31	82							
32	89							
33	51							

Figure 4.3.

By default, the numbers on the horizontal axis are the upper limits used to generate the frequency distribution. This may be **misleading** because these values can be confused with the midpoints of the intervals. To correct this misleading labeling, we have two possible alternatives.

One alternative is to replace the labels of the bars with the ***midpoints*** of the class intervals. The midpoints are the average of the upper and lower limit of each interval (see Figure 4.3). First, we create a column with the actual values of the midpoints.

- Enter in cell E2 a formula to compute the average of the upper and lower limit for the first class interval (i.e., (C2+D2)/2) and drag this function to the other entries in the column.
- Now, click with the mouse right button on the histogram and select the "Select Data:" option. On the *Horizontal (Category) Axis Labels* window click on Edit. A window opens to select the range for the axis label range. Select the range with the midpoints and click OK. Now the graph will have the correct values of the midpoints in the middle of the bars.

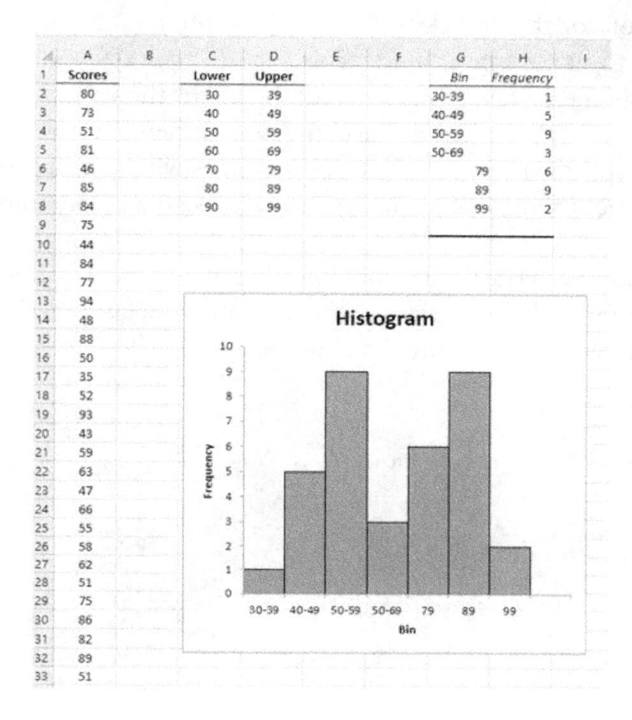

	A	C	D	G	H
1	Scores	Lower	Upper	Bin	Frequency
2	80	30	39	30-39	1
3	73	40	49	40-49	5
4	51	50	59	50-59	9
5	81	60	69	50-69	3
6	46	70	79	79	6
7	85	80	89	89	9
8	84	90	99	99	2
9	75				
10	44				
11	84				
12	77				
13	94				
14	48				
15	88				
16	50				
17	35				
18	52				
19	93				
20	43				
21	59				
22	63				
23	47				
24	66				
25	55				
26	58				
27	62				
28	51				
29	75				
30	86				
31	82				
32	89				
33	51				

Figure 4.4.

Another, simpler, alternative is to enter directly the class interval values in the Bin column of the frequency distribution table (see Figure 4.4). Because these values are linked to the labels on the axis of the histogram, the changes that we made to the bin values will appear also in the histogram plot. In the Excel worksheet 3 for this chapter, we start doing this change.

- Thus, starting with value 39 in the first entry, we type "30–39," then "40–49," and so on. The bin values for the first four intervals show the actual class interval. See plot on the right.

If we want a percentage histogram, we need to change the values on the vertical axis from frequency to percentage. In worksheet 4 of the Excel file for this chapter, we have the histogram with class intervals correctly labeled on the horizontal axis and frequencies on the vertical axis. To change the vertical axis to percentages we do the following (see Figure 4.5):

- First, we need a new column with the percentage of observations in each class interval. For the first class interval, put in I2, the expression =H2/35 to find the proportion for the first class interval (35 is the total frequency) and drag the function to all the entries in the table.

- To change the results to percentages, highlight the values in column I. Click on the Home tab in the top Excel menu and then click on the % icon on the Number sub-menu on the top bar. The values will be changed to percentages. To add one decimal, click on the Increase Decimal icon among the Number options.
- We replace the vertical axis with the values in the column I. Right click on the vertical axis of the histogram and click on the "Select Data" option. This will open the "Select Data Source" menu.

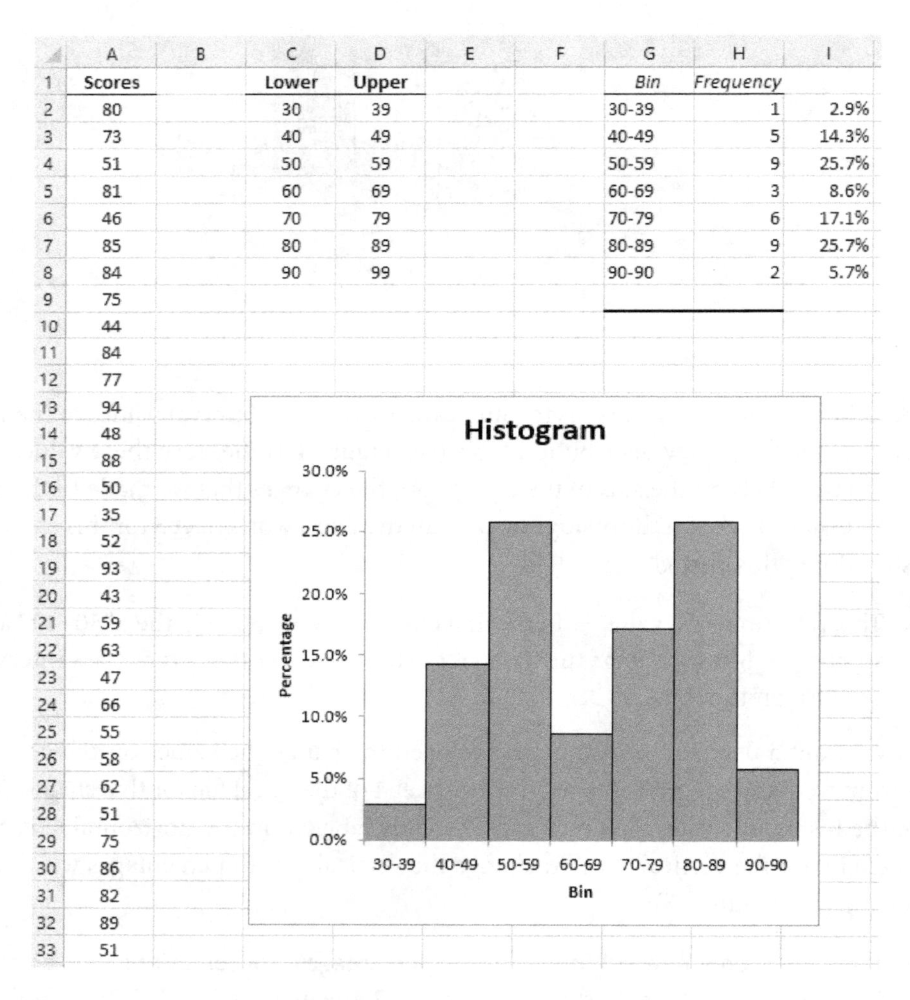

	A	B	C	D	E	F	G	H	I
1	Scores		Lower	Upper			Bin	Frequency	
2	80		30	39			30-39	1	2.9%
3	73		40	49			40-49	5	14.3%
4	51		50	59			50-59	9	25.7%
5	81		60	69			60-69	3	8.6%
6	46		70	79			70-79	6	17.1%
7	85		80	89			80-89	9	25.7%
8	84		90	99			90-90	2	5.7%
9	75								
10	44								
11	84								
12	77								
13	94								
14	48								
15	88								
16	50								
17	35								
18	52								
19	93								
20	43								
21	59								
22	63								
23	47								
24	66								
25	55								
26	58								
27	62								
28	51								
29	75								
30	86								
31	82								
32	89								
33	51								

Figure 4.5.

- In the "Legends Entries (Series)" select the "Frequency" entry and click Edit. In the "Series Values" box, replace the range of the frequencies (they are in column H) with the range of the percentages (in column I); we can highlight the cells with those values. Click enter, and then Ok and Ok. The vertical axis values change to percentages. Click OK and then change the label of the vertical axis to Percentages.

Histogram Using the Graph-Chart

Among the Charts options in the "Insert" Excel top menu, the option Histogram provides the easiest way to produce histograms in Excel. In worksheet 5 in the Excel file for this lesson we have again the scores for the 35 students, but now sorted from the smallest to the largest value. By the way, to sort a range of values we do the following:

- Select the range of values to sort.
- Go to the "Data" tab in the Excel top menu, and click on the sort icon. The default is to sort the values from smallest to largest.

To create a histogram using the Chart options, first we have to select a data range: in our case A2:A36.

- Click on the "Insert" tab in the Excel top menu. In the Charts menu click on the icon with blue bars (the Statistical charts), then select the first Histogram icon.

Excel will create a default histogram (see Figure 4.6) where:

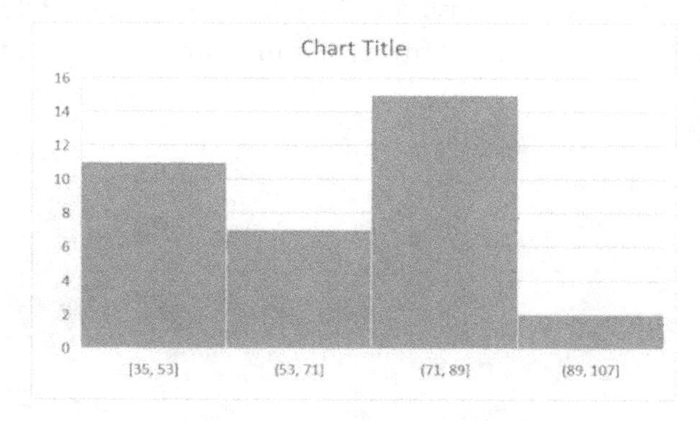

Figure 4.6.

(a) The default interval width is $\frac{3.5 * \text{SD}}{\sqrt[3]{n}}$, where SD is the standard deviation of the data and n is the number of values in the data. In the case of the data in this example, the width is 18 and 4 bars are displayed. We can later change this interval width at will.

(b) The default lower limit of the first class interval is the minimum value in the data; in our data, it is 35. We cannot change this option in Excel.

- To change the interval width, or the number of intervals, click twice on the horizontal axis to open the "Format Axis" menu. In the Axis options, we can enter in the "Bin width" box a value for another width, or we can enter in the "Number of bins" box a value for the number of bins. In this case, we select the option of "Bin width" equal to 10.
- To add the actual count of values s on top of the bars, first click on the graph. "Plus" and a "Paintbrush" icons will appear on the side of the group. Click on the plus box and select the option "Data Labels." The frequency for each interval will appear at the top of the bars (see Figure 4.7).

Notice that the bars axis labels are displayed as "open-close" class intervals. For example, the interval (55, 65) means that values ***larger than*** 55 and ***smaller or equal*** to 65 should be included in this interval. The only exception is the lowest interval (the one that starts with the minimum value in the data set); this is a "close-close" interval, i.e., [35, 45] that includes values ***larger or equal*** to 35 and ***smaller or equal*** to 45.

Even when we decide to use the same interval width, the shape of a histogram may differ when using the Histogram chart and the Histogram option in the Data Analysis ToolPak. This is because the Data Analysis ToolPak allows us to start the lower interval at a value different from the minimum, e.g., we used 30.

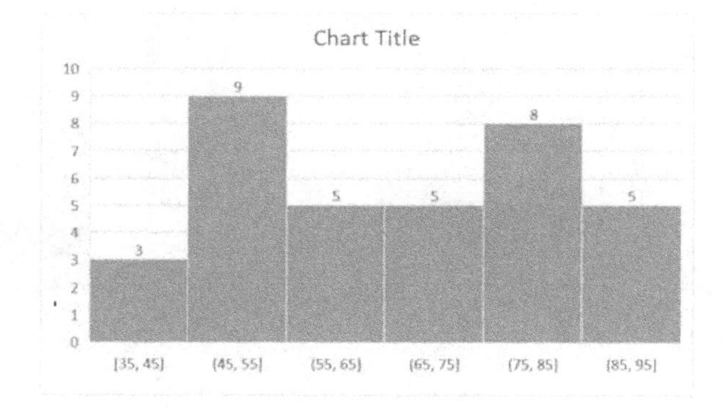

Figure 4.7.

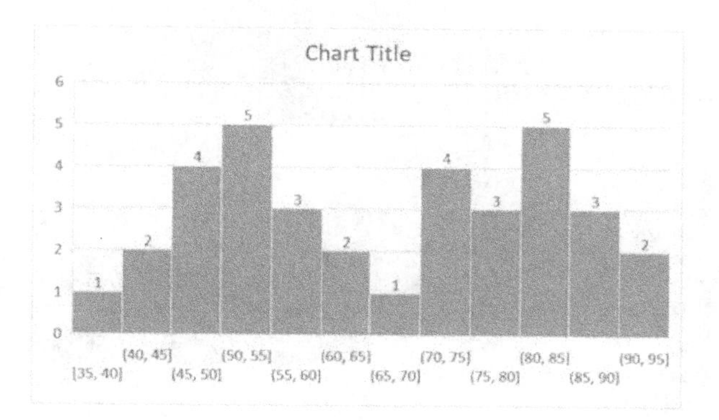

Figure 4.8.

We can easily try different interval widths using the Histogram chart. For example, if we change the interval width to 5, we will obtain a double peak shaped histogram, as in Figure 4.8.

Histograms Using Pivot Charts

We can use a pivot table as input to obtain the corresponding histogram and other graphs. The graphs or charts that use a pivot table as input, called pivot charts, are accessible through the Pivot Chart icon in the "Insert" tab of the top Excel menu.

For example, in worksheet 6 in the Excel file for this lesson we have again the students' score data. In addition, we have a pivot table for the data using an interval width of 5 points. We used the minimum value in the data, 35, to start the lower limit of the lower interval (see Figure 4.9).

- To add the histogram, click on any of the values in the first column of the pivot table and then on the "Insert" tab in the top Excel menu. In that menu, click on the Pivot Chart icon.
- An "Insert Chart" menu will appear. The default chart type option will be "Column." Select that one and click OK.
- The default chart will be similar to the one produced by the Data ToolPak. We have to eliminate the gaps between the bars, and erase the caption on the right-hand side as we described before. To add the data values on top of the columns, click on the bars and on the "plus" icon to add data labels.

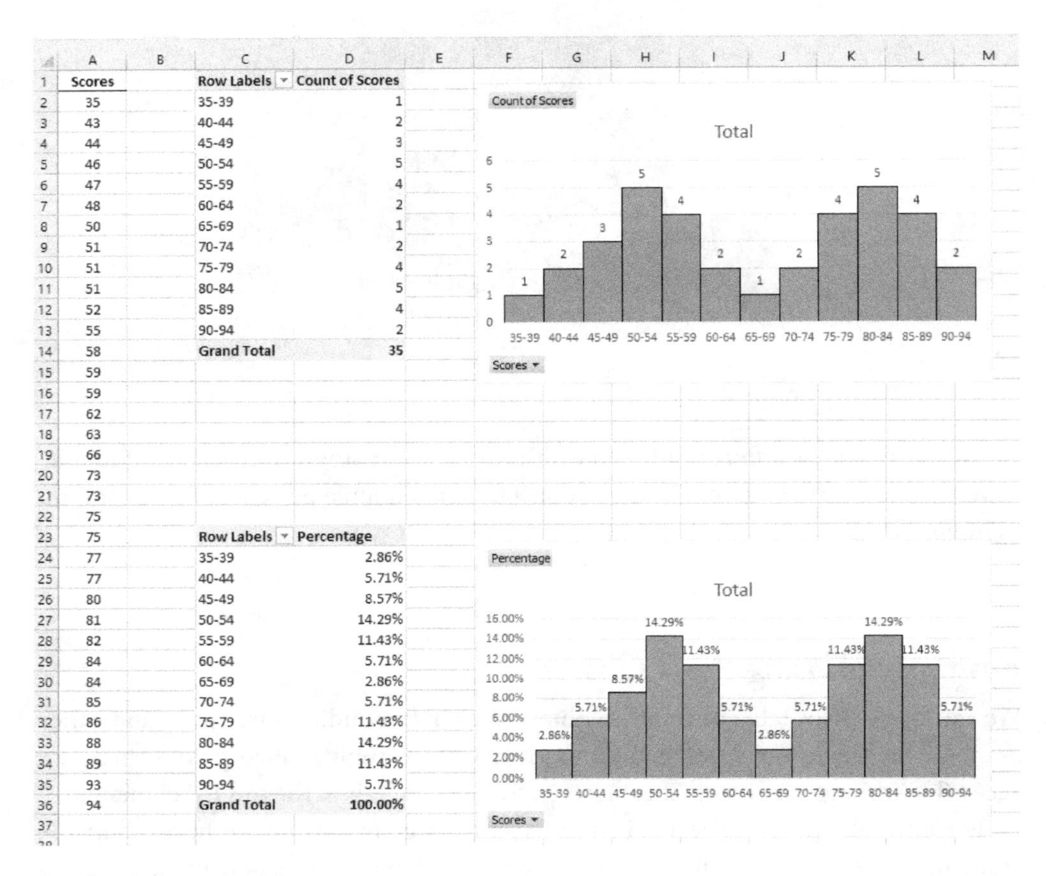

Scores
35
43
44
46
47
48
50
51
51
51
52
55
58
59
59
62
63
66
73
73
75
75
77
77
80
81
82
84
84
85
86
88
89
93
94

Row Labels	Count of Scores
35-39	1
40-44	2
45-49	3
50-54	5
55-59	4
60-64	2
65-69	1
70-74	2
75-79	4
80-84	5
85-89	4
90-94	2
Grand Total	**35**

Row Labels	Percentage
35-39	2.86%
40-44	5.71%
45-49	8.57%
50-54	14.29%
55-59	11.43%
60-64	5.71%
65-69	2.86%
70-74	5.71%
75-79	11.43%
80-84	14.29%
85-89	11.43%
90-94	5.71%
Grand Total	**100.00%**

Figure 4.9.

Notice that the histogram using Pivot Charts and the one using the Histogram chart do not produce exactly the same frequencies in each interval, even when both start at 35 and have the same width of 5. This is because the Pivot Chart uses close–close intervals, and the Histogram chart uses open–close intervals.

With pivot charts it is easy to change the vertical axis to percentages. In the same Excel worksheet 6 you will find another pivot chart with percentages instead of counts. We created that by simply clicking on the count column and changing the way the values are displayed to "% of Column total."

FREQUENCY POLYGONS USING PIVOT CHARTS

4.3

In worksheet 7 in the Excel file for this chapter, we copied again the scores for the 35 students and the pivot table for the frequency distribution with interval width 5. To add the frequency polygon for the data:

- Click on any of the values in the first column of the pivot table and then on the "Insert" tab in the top Excel menu. In that menu, click on the Pivot Chart icon.
- Change the default chart type of "Column" to "Line" and click OK. Excel will plot the frequency polygon using the midpoints of the class intervals (see the top graph in Figure 4.10).
- To change the vertical axis to percentages, simply click on the count column and change the way the values are displayed to "% of Column total" (see bottom graph in Figure 4.10).

CUMULATIVE POLYGON FOR PERCENTAGES

4.4

We can build cumulative percentage polygons using the Data ToolPak or a Pivot Chart.

Cumulative Percentage Polygon Using the Data ToolPak

In order to produce a cumulative percentage plot using the Histogram option of the Data Analysis ToolPak, we repeat the steps to generate a histogram, and, in addition, click the box option "Cumulative Percentage."

In worksheet 8 of the Excel file for this chapter, we have the students' score data and the lower and upper limits for the desired intervals.

- Click on the Data tab in the top Excel menu, click on the Data Analysis icon, and select the "Histogram" option.
- In the Histogram menu, enter the data range (A2:A36), the Bin range or upper limits (D2:D8), and as Output Range cell G1 (for the upper left corner of the table). Check the Chart Output box, ***and also the Cumulative Percentage box.*** Finally, press OK.

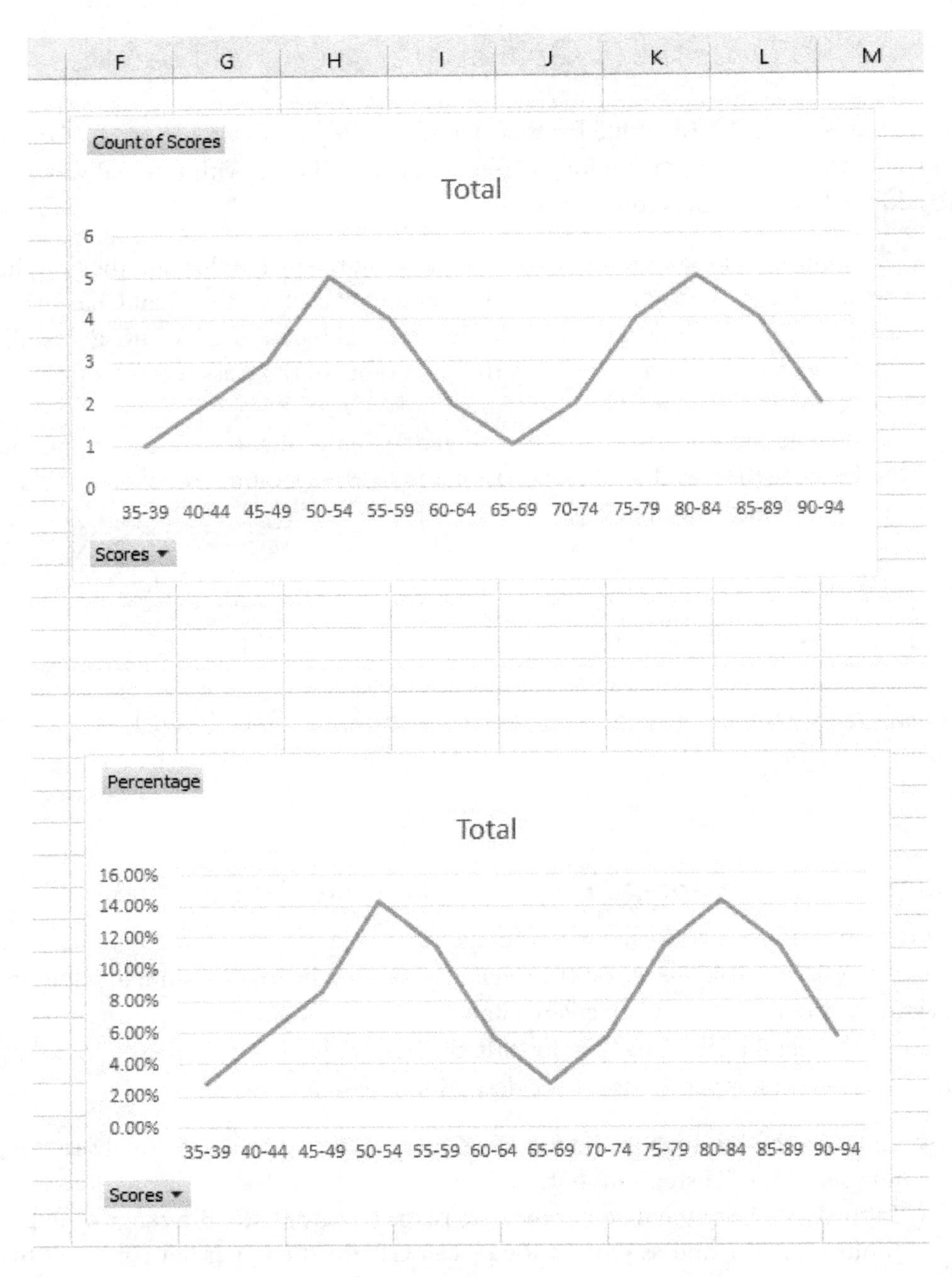
F
G
H
I
J
K
L
M
Count of Scores
Total
6
5
4
3
2
1
0
35-39 40-44 45-49 50-54 55-59 60-64 65-69 70-74 75-79 80-84 85-89 90-94
Scores
Percentage
Total
16.00%
14.00%
12.00%
10.00%
8.00%
6.00%
4.00%
2.00%
0.00%
35-39 40-44 45-49 50-54 55-59 60-64 65-69 70-74 75-79 80-84 85-89 90-94
Scores

Figure 4.10.

The frequency distribution table will have an additional column with the cumulative percentage for the data. In addition, the cumulative percentage polygon will overlay on top of the histogram.

As we did before for the histogram, we have to erase the series labels on the right (Frequency and Cumulative %) and eliminate the gaps between bars. In the column Bin, we edit the values by typing the actual class interval, i.e., instead of 39 we type in 30-39, etc. These new values will show on the horizontal axis of the plot. Finally, we rename the horizontal axis "Class Intervals," by clicking on the Bin label and type the new label. The resulting plot is in Figure 4.11.

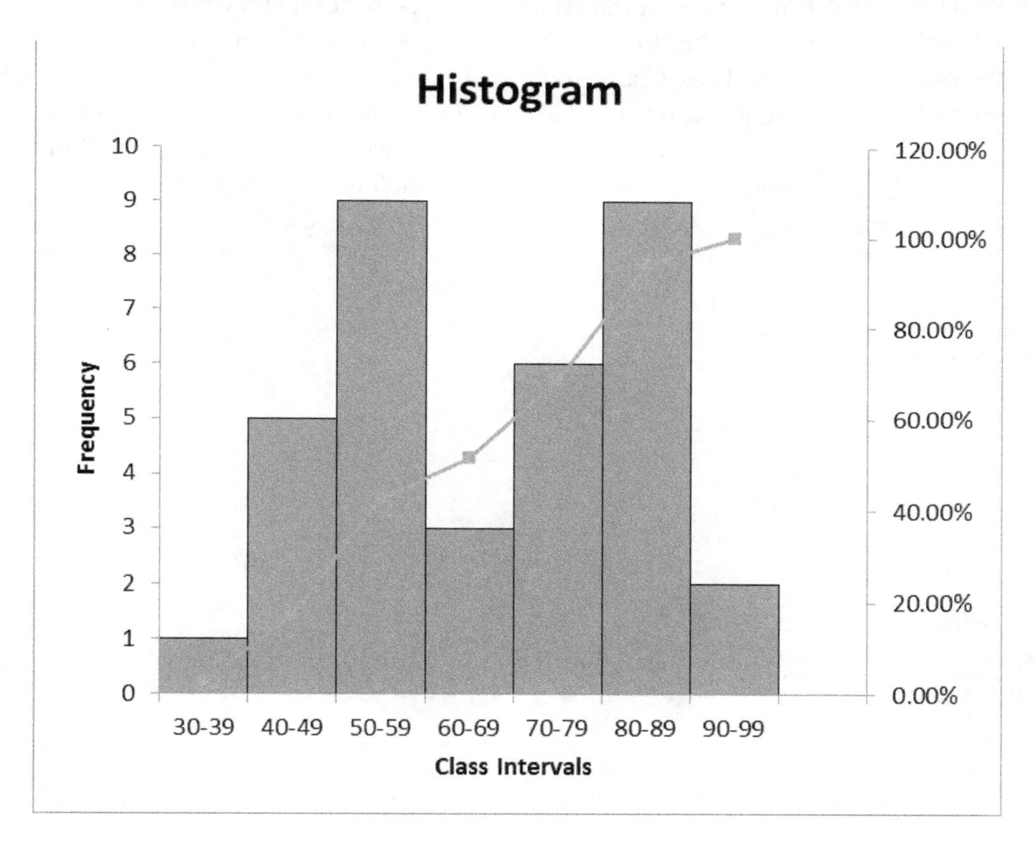

Figure 4.11.

Row Labels	Count of Scores
35-39	2.86%
40-44	8.57%
45-49	17.14%
50-54	31.43%
55-59	42.86%
60-64	48.57%
65-69	51.43%
70-74	57.14%
75-79	68.57%
80-84	82.86%
85-89	94.29%
90-94	100.00%
Grand Total	

Figure 4.12.

Cumulative Percentage Polygon Using Pivot Charts

In worksheet 9 in the Excel file for this lesson, we copied again the scores for the 35 students and the pivot table for the frequency distribution with interval width 5. However, in this case instead of counts or percentages, we have the "% Running Total In" as the entries in the second column (see Figure 4.12). We create the cumulative frequency polygon by inserting a line plot into the spreadsheet.

- Click on any of the values in the first column of the pivot table and then on the "Insert" tab in the top Excel menu. In that menu, click on the Pivot Chart icon.
- Change the default chart type of "Column" to "Line" and click OK. The cumulative frequency polygon will be plotted using the midpoints of the class intervals. The result is in Figure 4.13.

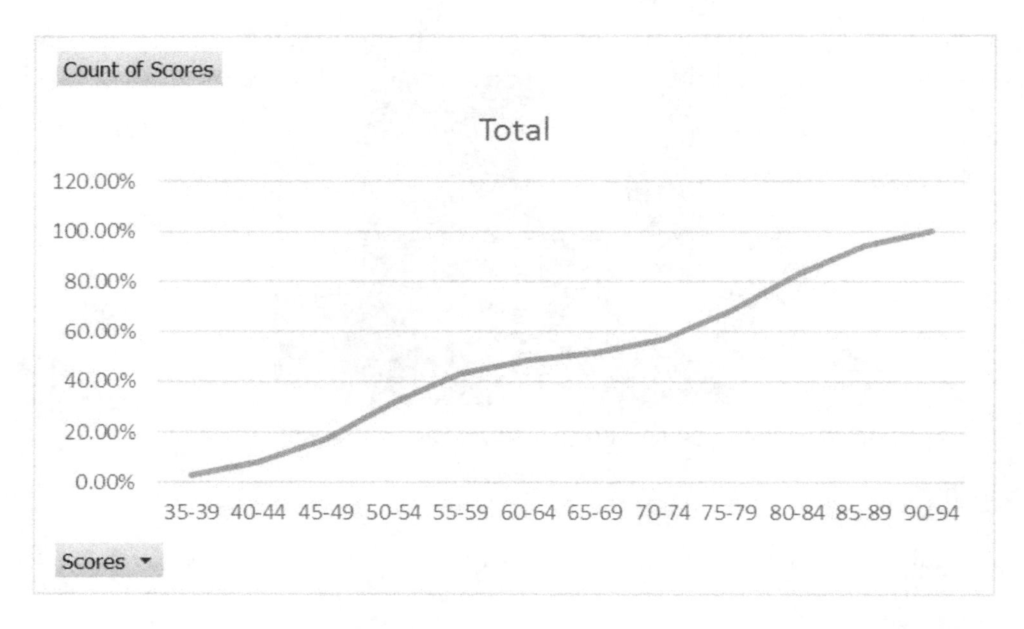

Figure 4.13.

PIE CHARTS

Histogram and frequency polygons are graphical representations for frequencies or cumulative frequencies of continuous variables. For data that is categorical, we only have frequencies or percentages of occurrence for each of the categories. The best plot for this type of data is the pie chart. In Excel, we can generate pie charts from raw data or from frequency tables that we can type in directly.

Pie charts for raw data are easy to obtain in Excel once we have a pivot table with the frequency distribution for the categorical data. In worksheet 10 in the Excel file for this chapter, you will find the raw data about the gender of 35 students. For this data we create a pivot chart for frequencies or counts of gender following the steps presented in the previous chapter. We create the pie chart by inserting a pie plot into the spreadsheet (see Figure 4.14).

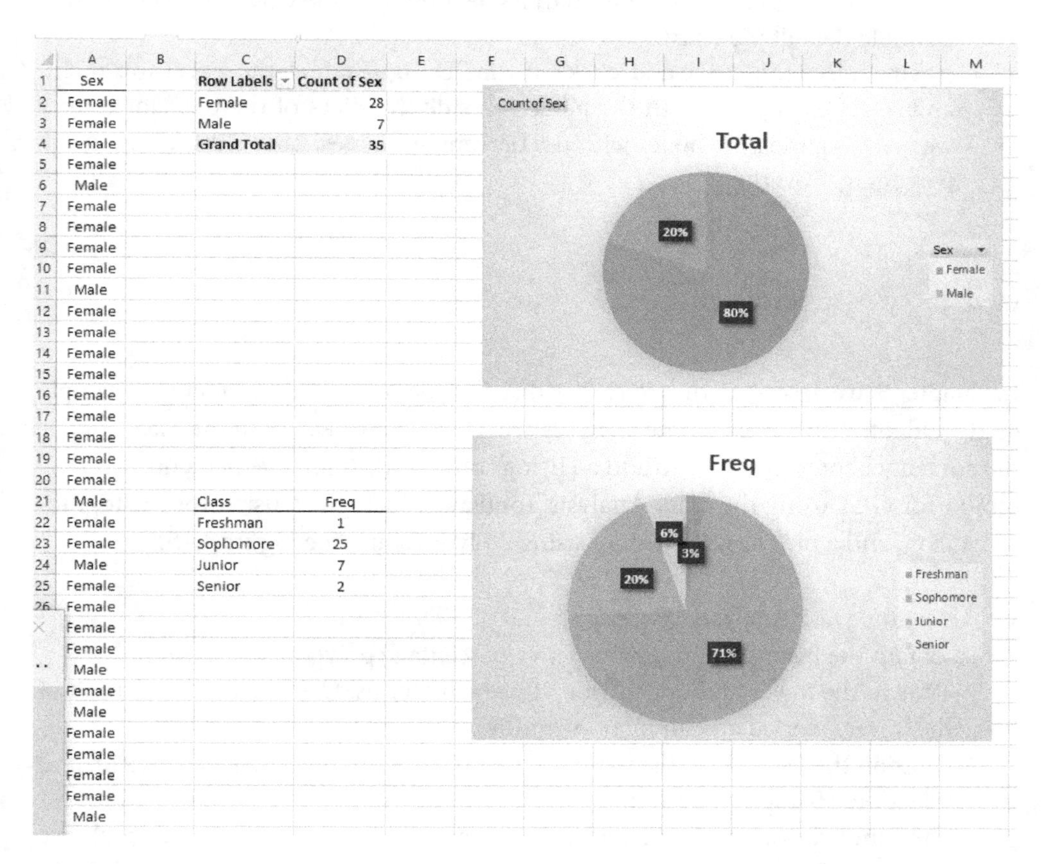

Figure 4.14.

- Click on any of the values in the first column of the pivot table and then on the "Insert" tab in the top Excel menu. In that menu, click on the Pivot Chart icon.
- Change the default chart type of "Column" to "Pie" and click OK.
- To add percentages to the chart, select the pie chart and click the "Design" tab option among the PivotChart Tools menu in the Excel top menu.
- Select among the design options the one that provides the percentage for each slice of the chart.

If we type directly in the frequency table for the categorical data, we can obtain the pie chart by using the Chart menu in the Insert tab. In the same worksheet 10, we type in a simple table of the college classification for the 35 students in a class.

- Highlight the values in the table, including the headers "Class" and "Freq." Select the INSERT tab in the top Excel menu and in the Charts menu select pie chart. From the list of possible charts, select the 2-D Pie Chart. Excel produces a simple default pie chart.
- Again, select the chart and click on the "Design" tab at the Excel top menu and select the option with the percent inside the slices of the pie. Notice that we can drag the little boxes with the percentage values. This is useful when the percentages overlap.

4.6 WORKED EXAMPLE

In Chapter 3, we described the frequency distribution of GPA and gender for a sample of 20 students. The results of the analyses are in the worksheet 12 in the Excel file for the current chapter. We want to add a Histogram and a cumulative percentage distribution for GPA using the Data Analysis ToolPak, a histogram using the Histogram chart icon, and a pie chart for gender using a Pivot Chart (see Figure 4.15).

1. Using the Data Analysis ToolPak.
   To obtain the histogram and cumulative percentage polygon for the GPA:
   - Type in the lower and upper limits for the intervals. Use 2 as the lower limit of the lower interval and an interval width of 0.3.
   - Click on the Data tab at the top Excel menu, click on the Data Analysis icon and select the Histogram option.
   - The input range is A2:A21 and the Bin range G2:G8. Check the "Cumulative Percentage" and the "Chart Output" boxes. For "Output range" select cell D10.

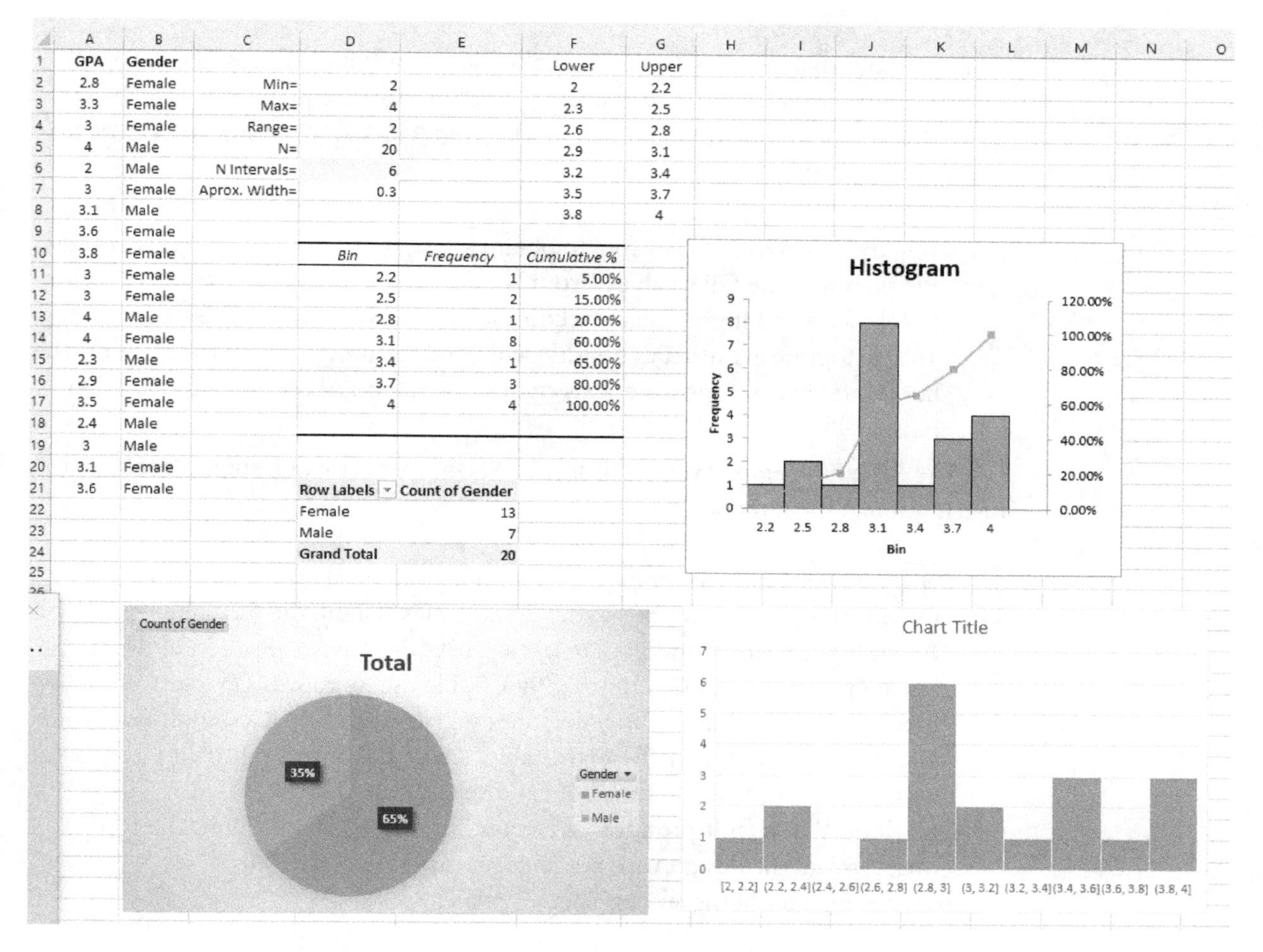

Figure 4.15.

- To edit the histogram, first, delete the "Frequency" and "Cumulative %" legends on the right-hand side of the plot.
- Second, eliminate the gaps between the bars. Click twice on the bars to open the Format Data Menu. In the Gap Width section, slide the marker to 0% gap.
- To add border' lines to the bars, click on the paint bucket icon and select the "Border" menu. In the Border menu, select solid line and as color black, then press close.
- You can make the plot bigger by dragging one of its corners. To add gridlines, click on the graph and then on the plus button, and check gridlines option.

Remember that the bins' labels are really the upper limits of the intervals.

The histogram seems to have a single "peak," or more frequent interval. The three "tallest" bars are at intervals 2.9–3.1, 3.8–4.0, and 3.5–3.7. The cumulative 50% falls in the interval 2.9–3.1

2. Using the Histogram Chart tool to obtain a histogram with 10 intervals or bins.
   - We highlight the GPA values (from A2 to A21), select the Insert tab at the top Excel menu and in the Charts section select the Histogram icon.
   - The histogram with default width will produce only three bars. Click on the horizontal axis to open the Axis option menu and select 10 as the number of bins.

With 10 bins, the interval width is 0.2. Again, there is still a single "peak" in the distribution at the interval (2.8, 3).

3. Using a Pivot Chart for gender distribution.
   - Obtain the pivot table for gender frequencies. Highlight the Gender column (include the header) and then insert a pivot table (use as location in the "Existing Worksheet" cell D21). In the Pivot Table menu, move "Gender to the Row box and to the Value box" (by default it will appear as the Count of Gender).
   - Highlight one entry in the pivot table, select the Insert tab in the top Excel menu, click on the Pie Chart icon, and select a pie chart.
   - Click on the default produced chart and, in the "Design" tab in the top Excel menu, select the design with percentages inside the chart slices.

   Sixty-five percent of the students were female.

A blank worksheet with only the data to repeat the analysis is in worksheet 11 of the Excel file for this chapter.

5 Measures of Central Tendency

M*easures of central tendency* are single values that describe the "typical" value in a data set. We have more than one way to define what a "typical" value is. For continuous data, the most commonly used measures of central tendency are:

The mean The sum of the values in the data set divided by the number of values. Or, the value from which the deviations of all the other values in the data set sum zero (function AVERAGE)
In summation notation: $\bar{X} = \sum_{i=1}^{n} X_i \Big/ N$

The median The value in the middle of the distribution of data values (function MEDIAN)

The mode The most frequent value in the data set (array function MODE.MULT)

Although the most used measure is the mean, there are situations in which the other measures are better in describing the data. When the values of the mean, median, and mode (assuming one mode only) are similar, the distribution of the data is likely to be ***symmetric***, i.e., its histogram peak around the value of the mean and the frequencies of other intervals getting lower in similar way when moving above or below the mean (see Figure 5.1). When the value of the mean is larger than the median, the data is said to be ***positively skewed***, i.e., its histogram has high frequencies at lower values with a long tail of low frequencies

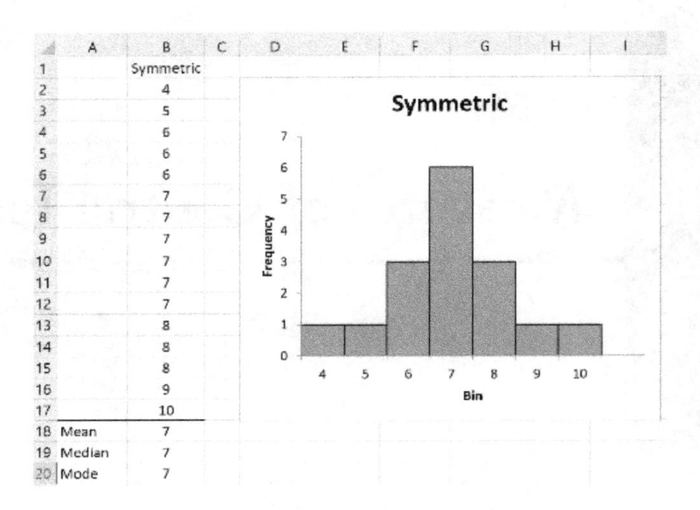

	A	B
1		Symmetric
2		4
3		5
4		6
5		6
6		6
7		7
8		7
9		7
10		7
11		7
12		7
13		8
14		8
15		8
16		9
17		10
18	Mean	7
19	Median	7
20	Mode	7

Figure 5.1.

at higher values (see Figure 5.2). When the value of the mean is smaller than the median, the data is said to be ***negatively skewed***, i.e., its histogram has high frequencies at higher values with a long tail of low frequencies at lover values (see Figure 5.3).

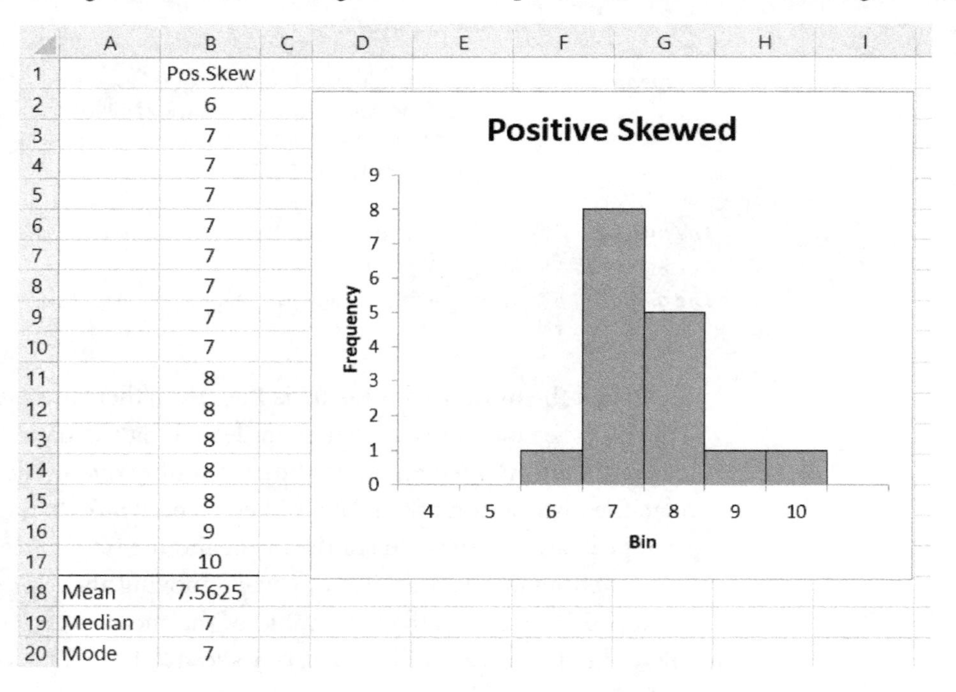

	A	B
1		Pos.Skew
2		6
3		7
4		7
5		7
6		7
7		7
8		7
9		7
10		7
11		8
12		8
13		8
14		8
15		8
16		9
17		10
18	Mean	7.5625
19	Median	7
20	Mode	7

Figure 5.2.

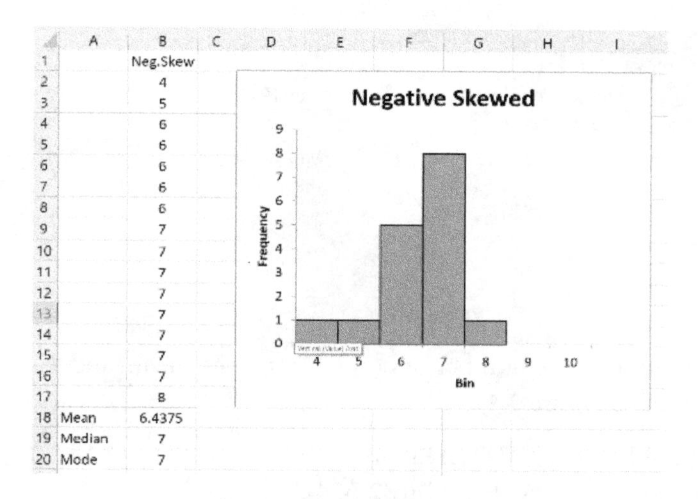

	A	B
1		Neg.Skew
2		4
3		5
4		6
5		6
6		6
7		6
8		6
9		7
10		7
11		7
12		7
13		7
14		7
15		7
16		7
17		8
18	Mean	6.4375
19	Median	7
20	Mode	7

Figure 5.3.

We use the measures of central tendency to answer questions such as:

- Is the typical household income better described by mean or median?
- What are the verbal achievements at different grades in elementary school?
- What is the most frequent number of internet-connected devices in a house?

OBJECTIVES

5.1

1. To compute measures of central tendency using Excel functions.
2. To use Excel Array Functions to compute multiple modes.
3. To find measures of central tendency using the Data Analysis ToolPak.
4. To use pivot tables to obtain means by groups of individuals.

MEASURES OF CENTRAL TENDENCY USING EXCEL FUNCTIONS

5.2

The file EXCEL COMPANION CHPT 5 CENTRAL TENDENCY contains information on GPA, ACT, and class year for a group of 72 undergraduate students. In the first worksheet of that Excel file, we describe the GPA scores in the range B2:B74.

	A	B	C	D	E	F	G
1	**Year**	**GPA:**	**ACT**		For GPA		
2	Freshman	2.9	20		N	72	
3	Freshman	3.8	23		Mean	3.294166667	
4	Junior	2.7	23		Median	3.3	
5	Junior	2.9	21				

Figure 5.4.

For GPA, we obtain the number of observations, the mean, and the median using Excel functions (see Figure 5.4):

- Type in cell F2: =COUNT(B2:B74).
- Type in cell F3: =AVERAGE(B2:B74).
- Type in cell F4: =MEDIAN(B2:B74).

Computing the mode introduces us to ***Excel array functions,*** i.e., functions whose output can be more than one value. Because a data set may have more than one mode, the function MODE.MULT returns not a single value but an *array* of possible values. To provide space for the possible values returned by the function, we first need to highlight a column of several cells for storing the results. Because we do not know how many modes are in a data set, we highlight a tentative number of cells (see Figure 5.5) in our example:

	A	B	C	D	E	F	G
1	**Year**	**GPA:**	**ACT**		For GPA		
2	Freshman	2.9	20		N	72	
3	Freshman	3.8	23		Mean	3.294166667	
4	Junior	2.7	23		Median	3.3	
5	Junior	2.9	21				
6	Junior	2.9	32		Modes	=MODE.MULT(B2:B73)	
7	Junior	2.96	34				
8	Junior	2.98	18				
9	Junior	2.99	20				
10	Junior	3	23				
11	Junior	3	29				
12	Junior	3.02	26				
13	Junior	3.1	24				
14	Junior	3.18	26				

Figure 5.5.

- Highlight six cells starting at cell F6.
- Once we highlight the cells, we enter =MODE.MULT(B2:B73). This function will appear in the top cell of the highlighted cells.
- To execute and array function, we press ***Ctrl+Shift+Enter.***

In this example, the mode function returns four modes: 3.8, 3, 3.4, and 4 (see Figure 5.6). These values occupy four of the highlighted cells. The remaining two cells were not necessary and they show a #N/A symbol.

Sometimes we want to know how many times the mode(s) appears in the data set. The function COUNTIF tallies the frequency of observations in a given range subject to a particular condition. For example, we would like to know how many observations in the range B2:B73 have the value of our modes. We start with the mode 3.8 in cell F6 (see Figure 5.6).

- In cell G6, we enter =COUNTIF(B2:B73, F6). In other words, we ask to count how many observations in the range B2:B73 have the value in cell F6. We can enter the ranges and cell by simply dragging or clicking on the corresponding cells. For 3.8 we will obtain 5, i.e., five students have a GPA of 3.8.
- Select cell G6 and use the lower right box to drag the function for the other modes. As we expect, all the other mode values (3, 3.4, and 4) also have the same frequency of five students (see Figure 5.7).

	A	B	C	D	E	F	G	H
1	Year	GPA:	ACT		For GPA			
2	Freshman	2.9	20		N	72		
3	Freshman	3.8	23		Mean	3.294166667		
4	Junior	2.7	23		Median	3.3		
5	Junior	2.9	21					
6	Junior	2.9	32		Modes	3.8	=COUNTIF(B2:B73, F6)	
7	Junior	2.96	34			3		
8	Junior	2.98	18			3.4		
9	Junior	2.99	20			4		
10	Junior	3	23			#N/A		
11	Junior	3	29			#N/A		
12	Junior	3.02	26					
13	Junior	3.1	24					
14	Junior	3.18	26					

Figure 5.6.

	A	B	C	D	E	F	G
1	Year	GPA:	ACT		For GPA		
2	Freshman	2.9	20		N	72	
3	Freshman	3.8	23		Mean	3.294166667	
4	Junior	2.7	23		Median	3.3	
5	Junior	2.9	21				
6	Junior	2.9	32		Modes	3.8	5
7	Junior	2.96	34			3	5
8	Junior	2.98	18			3.4	5
9	Junior	2.99	20			4	5
10	Junior	3	23			#N/A	
11	Junior	3	29			#N/A	
12	Junior	3.02	26				
13	Junior	3.1	24				
14	Junior	3.18	26				

Figure 5.7.

5.3 MEASURES OF CENTRAL TENDENCY USING THE DATA ANALYSIS TOOLPAK

We can obtain the measures of central tendency, and other descriptive statistics, using the Data Analysis ToolPak.

- Click the DATA tab in the top Excel menu and then click the Data Analysis icon.
- In the Data Analysis menu select "Descriptive Statistics" and then OK (see Figure 5.8).
- As input range select B2 to B73. Select "Output Range" and give as starting cell E15 (this will be the top-left corner of the table). Check the box for "Summary statistics" (see Figure 5.9). Then press OK.

Excel produces a table of descriptive statistics, i.e., mean, median, and mode. However, the table only reports one of the modes of the data (see Figure 5.10). In our case, we know that the data has four modes, but only the 3.8 mode is reported.

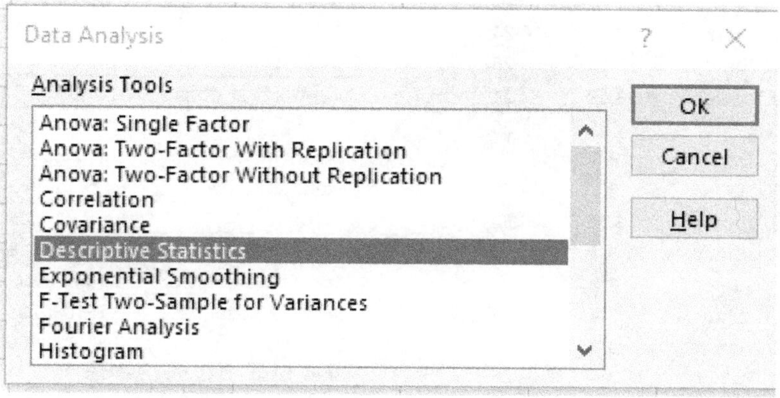

Data Analysis

Analysis Tools

Anova: Single Factor
Anova: Two-Factor With Replication
Anova: Two-Factor Without Replication
Correlation
Covariance
Descriptive Statistics
Exponential Smoothing
F-Test Two-Sample for Variances
Fourier Analysis
Histogram

OK
Cancel
Help

Figure 5.8.

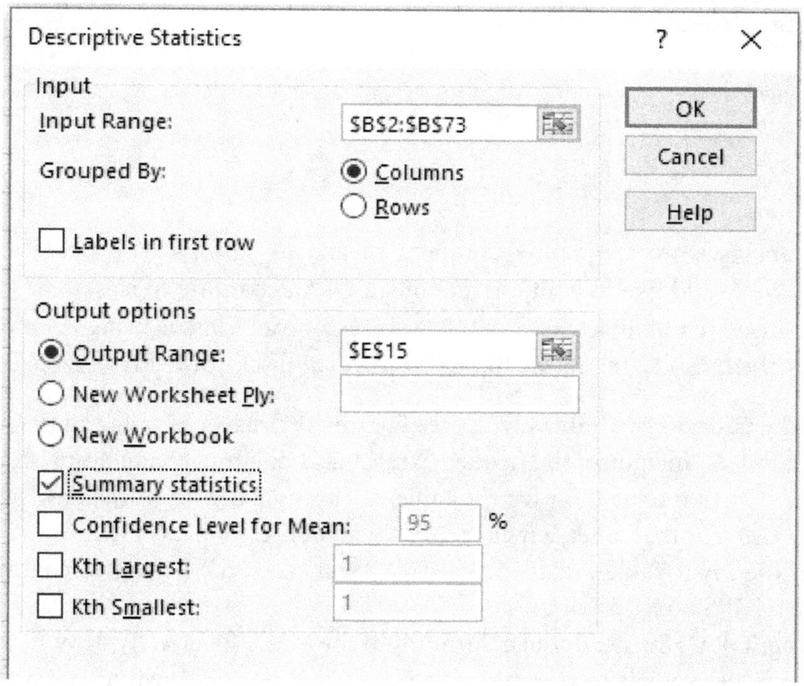

Descriptive Statistics

Input

Input Range: $B$2:$B$73

Grouped By: Columns / Rows

Labels in first row

Output options

Output Range: $E$15

New Worksheet Ply:

New Workbook

Summary statistics

Confidence Level for Mean: 95 %

Kth Largest: 1

Kth Smallest: 1

OK
Cancel
Help

Figure 5.9.

13	Junior	3.1	24			
14	Junior	3.18	26			
15	Junior	3.19	21		*Column1*	
16	Junior	3.2	26			
17	Junior	3.3	21		Mean	3.294166667
18	Junior	3.3	25		Standard Error	0.052417963
19	Junior	3.36	30		Median	3.3
20	Junior	3.4	19		Mode	3.8
21	Junior	3.4	26		Standard Deviation	0.444781162
22	Junior	3.4	27		Sample Variance	0.197830282
23	Junior	3.6	22		Kurtosis	-0.715323066
24	Junior	3.6	23		Skewness	-0.182361134
25	Junior	3.6	28		Range	1.7
26	Junior	3.77	32		Minimum	2.3
27	Junior	3.8	17		Maximum	4
28	Junior	3.8	24		Sum	237.18
29	Junior	3.82	25		Count	72
30	Senior	2.3	21			

Figure 5.10.

5.4 MEANS BY CATEGORIES USING PIVOT TABLES

In data analysis, we frequently compute means by groups of observations. For example, we would like to compute the mean GPA according to class year. In Excel, the easiest way to obtain a table with the means for each class is using pivot tables.

Check the INSERT tab in the top Excel menu and select the Pivot Table icon.

- In the Create Pivot Table menu (see Figure 5.11) select as range all the values in column A (including the header "Year") and column B (including the header "GPA"). As the location for the table, click on "Existing worksheet" and enter cell E33 (for the upper-left corner of the table). Then press OK.
- On the Pivot Table Fields menu, drag "Year" (i.e., the header of column A) to the ROWS box.
- Drag "GPA" (the header of column B) to the VALUES box. By default, the function Excel applies to the values is Sum. Click on the down arrow next to the "Sum of GPA" in the VALUES box and select "Value Field Settings." In the menu select "Average" and then OK.

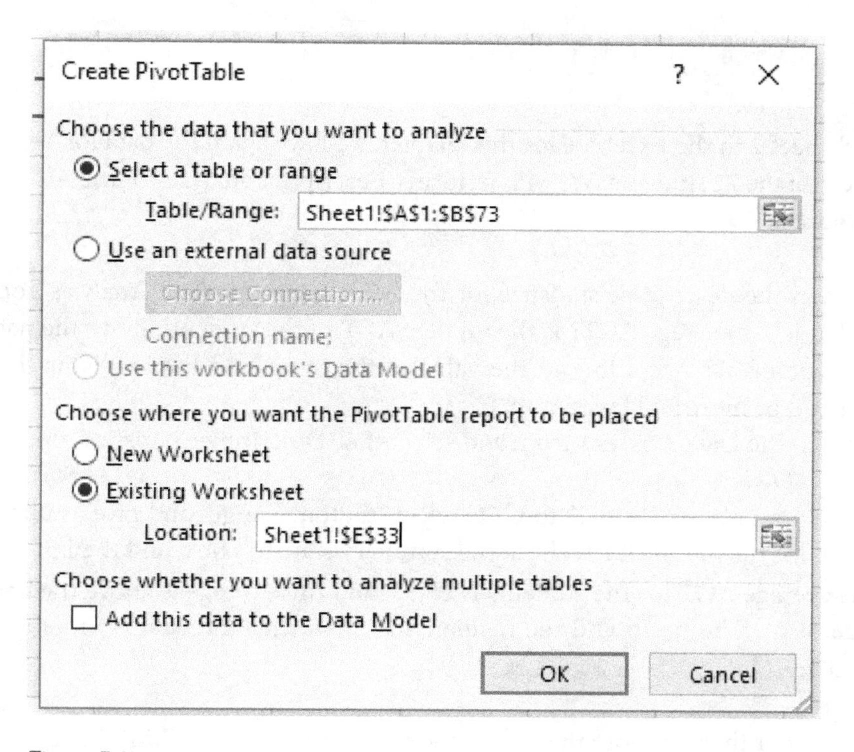

Figure 5.11.

- Drag again “GPA” to the “VALUES” box. Click again on the down arrow next to the GPA, select “Value Field Settings” and change the function to apply to “Count.”

The table for means and counts appear in the desired location (see Figure 5.12). We can see that the smallest average GPA corresponds to the eight seniors. The average GPA for freshman, 3.35, is computed using only two observations.

Row Labels	Average of GPA:	Count of GPA:
Freshman	3.35	2
Junior	3.279615385	26
Senior	3.12125	8
Sophomore	3.34	36
Grand Total	**3.294166667**	**72**

Figure 5.12.

5.5 WORKED EXAMPLE

In worksheet 2 in the Excel file for this chapter, we have again the data for Year, GPA and ACT of the 72 students. We will perform a descriptive analysis for the ACT values (see Figure 5.13).

1. Obtain the descriptive statistics for the ACT using the Data Analysis ToolPak. What is the average ACT? Is the average ACT of the students above the national average of 21? According to the value of the mean and the media, is the data skewed or more and less symmetric?
   - Click the Data Analysis icon, and select the "Descriptive Statistics" option and then OK.
   - As input range select C2 to C74. Select "Output Range" and give as upper-left cell for the table cell E1, check the summary statistics box, and then press OK.

   The average ACT for the students is 25.33, and this average is above the national average of 21. The mean and the median are close, thus the data is more and less symmetric.

2. Explore if there is more than one mode by using the MODE.MULT array function. How many modes are there in the data? Report these modes. What is the frequency of these modes?
   - We highlight six cells, starting with F18. We enter in F18 the function = MODE.MULT(C2:C7) and press ***Ctrl+Shift+Enter***.
   - To find the frequency of the modes, enter in G18 the function = COUNTIF(C2:C73,"23").

   There are two modes: 23 and 26. Each one occurs nine times.

3. Obtain the average ACT and count accordingly class year using pivot tables. Which class has the higher ACT? How many students are in that class? Which class has the lowest ACT? How many students are in that class?
   - Click on the Pivot Table icon. In the Create Pivot Table menu select as range all the three columns with their headers. As the location for the table, select "Existing worksheet" and enter cell E27 for the table upper-left corner, then press OK.

	A	B	C	D	E	F	G	H
1	**Year**	**GPA:**	**ACT**		*ACT*			
2	Freshman	2.9	20					
3	Freshman	3.8	23		Mean	25.33333333		
4	Junior	2.7	23		Standard Error	0.475124562		
5	Junior	2.9	21		Median	25.5		
6	Junior	2.9	32		Mode	23		
7	Junior	2.96	34		Standard Devi	4.031565593		
8	Junior	2.98	18		Sample Variar	16.25352113		
9	Junior	2.99	20		Kurtosis	-0.440345403		
10	Junior	3	23		Skewness	0.069272604		
11	Junior	3	29		Range	17		
12	Junior	3.02	26		Minimum	17		
13	Junior	3.1	24		Maximum	34		
14	Junior	3.18	26		Sum	1824		
15	Junior	3.19	21		Count	72		
16	Junior	3.2	26					
17	Junior	3.3	21					
18	Junior	3.3	25		MODES	23	9	
19	Junior	3.36	30			26		
20	Junior	3.4	19			#N/A		
21	Junior	3.4	26			#N/A		
22	Junior	3.4	27			#N/A		
23	Junior	3.6	22			#N/A		
24	Junior	3.6	23			#N/A		
25	Junior	3.6	28					
26	Junior	3.77	32					
27	Junior	3.8	17		**Row Labels**	**Average of ACT**	**Count of ACT**	
28	Junior	3.8	24		Freshman	21.5	2	
29	Junior	3.82	25		Junior	24.69230769	26	
30	Senior	2.3	21		Senior	25.25	8	
31	Senior	2.4	27		Sophomore	26.02777778	36	
32	Senior	2.5	22		**Grand Total**	**25.33333333**	**72**	
33	Senior	3	23					

Figure 5.13.

- On the Pivot Table Fields menu, drag "Year" (i.e., the header of column A) to the ROWS box. Then, drag "ACT" (the header of column C) to the VALUES box. Click on the down arrow point next to the Sum label and select "Value Field Settings." In the menu, select "Average" and then OK. Drag again "ACT" to the "VALUES" box. Now, change the function to apply to the values to "Count."

The 36 sophomores have the largest average ACT, 26.03. The two freshmen have the lowest average ACT: 21.5.

The results of these analyses are in the worksheet 3 in the Excel file for this chapter.

6 Measures of Variability

Measures of variability are values that provide information about the diversity or spread of the values in a data set. The most commonly used measure of variability, the ***standard deviation***, expresses the dispersion of the values in a data set around their mean. The standard deviation, as the mean, are in the original units of the data. Thus, if we describe the height of female students in inches, as "*a mean of 62 with standard deviation 2,*" we refer that the average height was 62 inches, and the variability around this values is 2 inches. Standard deviation tells us about the variability in the data. For example, suppose that the height of the female high school basketball team A has a mean of 68.2 and standard deviation 2.1, while for basketball team B the mean is 69.1 and the standard deviation is 1.2. We can say that team B is not only taller than team A, but also that players are more similar in their heights (i.e., less variable) than team A.

The standard deviation is a function of two other quantities that are also used to describe variability in data: the ***variance*** and the sum of squared deviations or simply ***sum of squares***. When dealing with a sample of values, these measures are:

The sample standard deviation, S: The (positive) squared root of the variance (function STDEV.S)

$$S = \sqrt{S^2}$$

The sample variance, S^2: The "average" squared deviations of the values from the mean (function VAR.S):

$$S^2 = \sum_{i=1}^{N}(X_i - \bar{X})^2 \Big/ (N-1) = SS/(N-1)$$

The sum of squared deviations, SS: The sum of the squared deviations from the mean:

$$SS = \sum_{i=1}^{N}(X_i - \bar{X})^2$$

Excel has a function to compute SS directly (function DEVSQ), but we can easily obtain this value from the variance and the number of observations in the data set:

$$SS = (N-1)S^2$$

We will see later that measures of variability, such as variance, are not only important in describing the data, but also in testing for differences among measures of central tendency.

Other indicators of variability in the data do not refer directly to the average of the values. The ***five-number summary*** and its corresponding graph, the ***boxplot***, use values that define the 25%, 50%, and 75% cumulative percentage of the distribution to describe and graph the variability of the data. We use measures of variability to answers questions such as:

- Which football kicker is more consistent in the length of his kick?
- We have the ACT means for two groups, so can we use their variability to know which mean better represents the group?
- What type of stocks is more volatile than other type of stocks?

6.1 OBJECTIVES

1. To compute measures of variability around the mean using Excel functions.
2. To obtain five-number summary and boxplots for a data set.
3. To obtain boxplots by group of individuals.
4. To use pivot tables to obtain measures of variability by group of individuals.

MEASURES OF DISPERSION USING EXCEL FUNCTIONS

6.2

In the first worksheet of the file EXCEL COMPANION CHPT 6 MEASURES OF VARIABILITY.XLSX, we have the class year (in column A), GPA (in column B), and ACT (in column C) for 72 undergraduate students.

We compute the number of observations in the data set, the sample variance, and the standard deviation of GPA using the respective Excel functions (see Figure 6.1):

- In F2, we type =COUNT(B2:B73).
- In F3, we write =VAR.S(B2:B73).
- And, in F4 we enter =STDEV.S(B2:B73).

	A	B	C	D	E	F	G
1	Year	GPA:	ACT		For GPA		
2	Freshman	2.9	20		N	72	
3	Freshman	3.8	23		Variance	0.197830282	
4	Junior	2.7	23		Stnd. Deviation	0.444781162	
5	Junior	2.9	21				
6	Junior	2.9	32		SS	14.04595	
7	Junior	2.96	34				
8	Junior	2.98	18				
9	Junior	2.99	20		*Column1*		
10	Junior	3	23				
11	Junior	3	29		Mean	3.294166667	
12	Junior	3.02	26		Standard Error	0.052417963	
13	Junior	3.1	24		Median	3.3	
14	Junior	3.18	26		Mode	3.8	
15	Junior	3.19	21		Standard Deviation	0.444781162	
16	Junior	3.2	26		Sample Variance	0.197830282	
17	Junior	3.3	21		Kurtosis	-0.715323066	
18	Junior	3.3	25		Skewness	-0.182361134	
19	Junior	3.36	30		Range	1.7	
20	Junior	3.4	19		Minimum	2.3	
21	Junior	3.4	26		Maximum	4	
22	Junior	3.4	27		Sum	237.18	
23	Junior	3.6	22		Count	72	
24	Junior	3.6	23				

Figure 6.1.

To compute the sum of squared deviations (SS), we multiply the variance by (N – 1) or we can use the function DEVSQ(B2:B27):

- In our example, the variance is in F3 and the count in F2, so then in F6 type= (F2-1)*F3.

6.3 MEASURES OF VARIABILITY USING THE DATA ANALYSIS TOOLPAK

As we saw in the previous chapter, we can get several descriptive statistics, including the variance and standard deviation, by using the descriptive statistics option in the Data Analysis ToolPak.

- Click on Data tab in the top Excel menu and then on the Data Analysis icon. Select the "Descriptive Statistics" options.
- As Input range select B2 to B73. Select "Output Range" and give as starting cell E9, and check summary statistics. Then press OK (see output also in Figure 6.1).

6.4 QUARTILES, INTERQUARTILE RANGE (IQR), AND FIVE-NUMBER SUMMARY

The ***five-number summary*** is another way to describe the variability or spread in the data. The numbers in the five-number summary are: the minimum, the ***first quartile***, the median, the ***third quartile***, and the maximum of the data.

The ***first quartile*** (usually represented as Q1) is the value that marks the lowest 25% of the data distribution, and the ***third quartile*** (or Q3) is the value marking the upper 25% of the distribution (or the lowest 75%).

The Descriptive Statistics option of the Data Analysis ToolPak outputs the minimum, maximum, and median of the data, but not the quartiles, which we can obtain by using the QUARTILE function. For example, to obtain the five-number summary for GPA, we use the following functions (see Figure 6.2):

- In F26 to obtain the minimum type: =MIN(B2:B74)
- In F27 to obtain the first quartile type: =QUARTILE(B2:B74, 1)

9	Junior	2.99	20		Column1	
10	Junior	3	23			
11	Junior	3	29		Mean	3.294166667
12	Junior	3.02	26		Standard Error	0.052417963
13	Junior	3.1	24		Median	3.3
14	Junior	3.18	26		Mode	3.8
15	Junior	3.19	21		Standard Deviation	0.444781162
16	Junior	3.2	26		Sample Variance	0.197830282
17	Junior	3.3	21		Kurtosis	-0.715323066
18	Junior	3.3	25		Skewness	-0.182361134
19	Junior	3.36	30		Range	1.7
20	Junior	3.4	19		Minimum	2.3
21	Junior	3.4	26		Maximum	4
22	Junior	3.4	27		Sum	237.18
23	Junior	3.6	22		Count	72
24	Junior	3.6	23			
25	Junior	3.6	28			
26	Junior	3.77	32		Minimum	2.3
27	Junior	3.8	17		Q1	3
28	Junior	3.8	24		Median	3.3
29	Junior	3.82	25		Q3	3.7
30	Senior	2.3	21		Maximum	4
31	Senior	2.4	27		IQR	0.7

Figure 6.2.

- In F28 to obtain the median type: =MEDIAN(B2:B74)
- In F29 to obtain the third quartile type: =QUARTILE(B2:B74, 3)
- In F30 to obtain the maximum type: =MAX(B2:B74)

Notice that options 1 and 3 in the QUARTILE functions refer to the first and third quartiles. Replacing that value with 2 will produce the median.

The ***interquartile range, IQR***, is the difference between the third and fourth quartiles.

- In F31 to obtain the IQR type: = F29-F27

6.5 BOXPLOT

A boxplot is the graphical representation of the five-number summary. Excel versions after 2016 provide among the statistical charts an option to obtain boxplot. To obtain the boxplot for GPA (see Figure 6.3):

- First, highlight the GPA data. Click the "Insert" tab in the Excel top menu.
- In the Charts menu click on the icon for Statistical charts (the icon is a histogram).
- From the drop-down menu, select "Box and whisker plot."

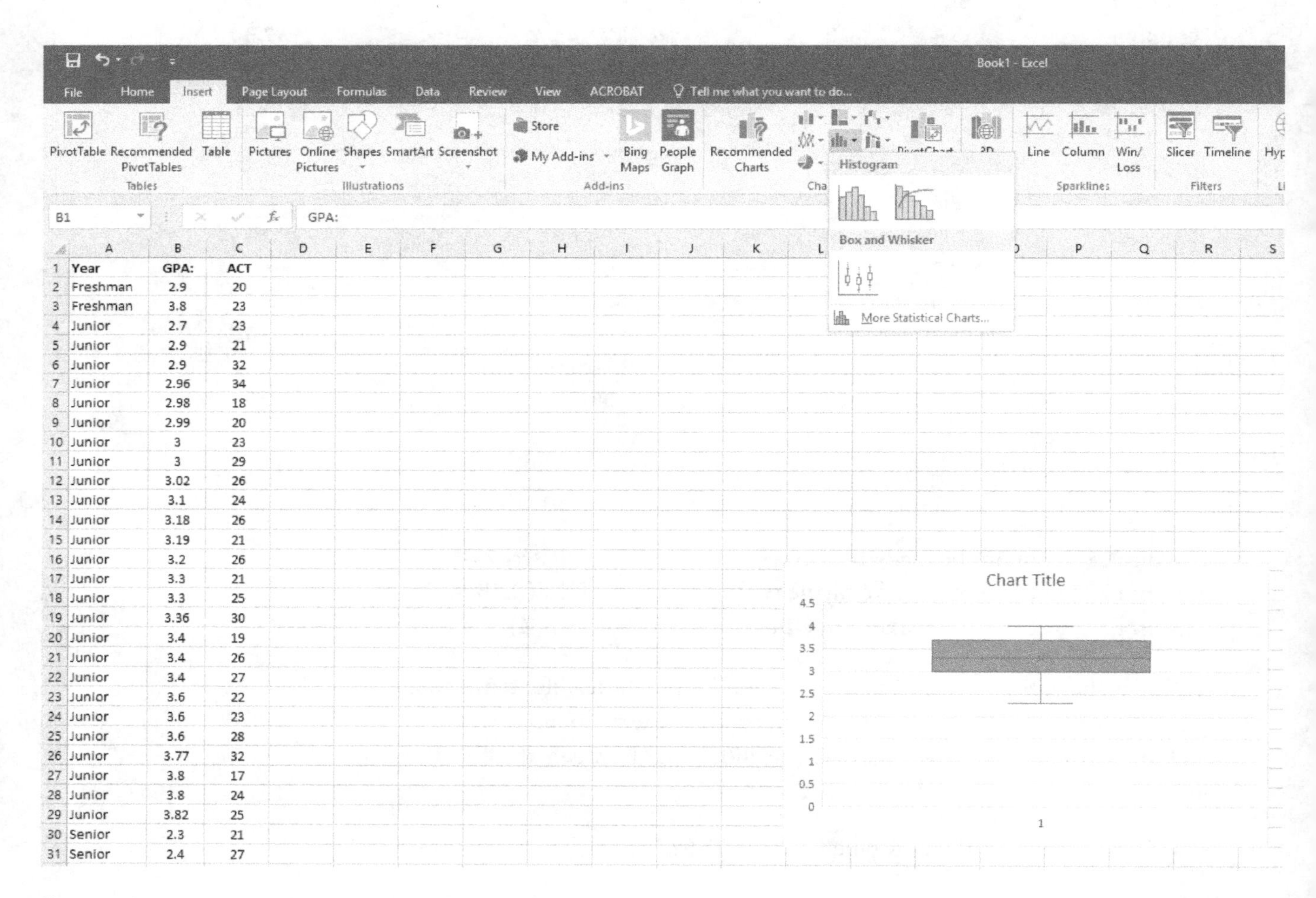

	A	B	C
1	Year	GPA:	ACT
2	Freshman	2.9	20
3	Freshman	3.8	23
4	Junior	2.7	23
5	Junior	2.9	21
6	Junior	2.9	32
7	Junior	2.96	34
8	Junior	2.98	18
9	Junior	2.99	20
10	Junior	3	23
11	Junior	3	29
12	Junior	3.02	26
13	Junior	3.1	24
14	Junior	3.18	26
15	Junior	3.19	21
16	Junior	3.2	26
17	Junior	3.3	21
18	Junior	3.3	25
19	Junior	3.36	30
20	Junior	3.4	19
21	Junior	3.4	26
22	Junior	3.4	27
23	Junior	3.6	22
24	Junior	3.6	23
25	Junior	3.6	28
26	Junior	3.77	32
27	Junior	3.8	17
28	Junior	3.8	24
29	Junior	3.82	25
30	Senior	2.3	21
31	Senior	2.4	27

Figure 6.3.

We can edit or change the look of the default boxplot. In our example:

- To start the vertical axis at 2.0, click twice on the vertical axis of the plot. In the "Axis option" menu, enter 2 for the minimum value.
- Second, to eliminate the filling of the box, click twice on the plot. Select the paint bucket icon. Select the "Fill" menu and check the "No fill option."
- Third, to eliminate the *X* inside the box (it is the marker for the mean of the data), click twice on the *X* inside the box, and in the Format Data Point menu uncheck the "Show mean markers" box.
- Finally, to display the five-number summary on the plot, click on the plot and on the plus button that will appear on the top of the plot. Select the option "Data labels" (see Figure 6.4).

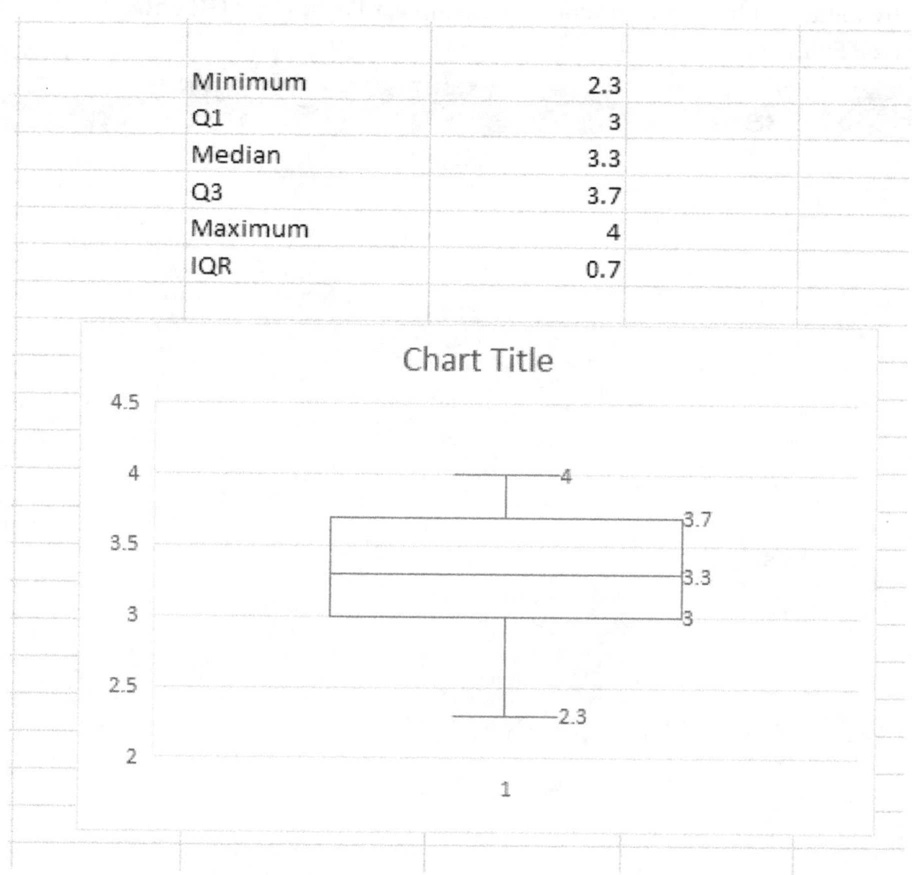

Minimum	2.3
Q1	3
Median	3.3
Q3	3.7
Maximum	4
IQR	0.7

Figure 6.4.

Notice that the values defining the "box" in the plot are the first quartile, the median, and the third quartile. This plot does not have ***outliers***. Outliers in a boxplot are observations that are larger than Q3+1.5 IQR or smaller than Q1-1.5IQR. When there are outliers in a data set, each one is represented by an individual marker in the boxplot. The "whiskers" in a boxplot end in the maximum and minimum values that are NOT outliers.

Boxplot by Group of Observations

The boxplot chart in Excel allows us to produce boxplots by groups of observations that are identified by a text variable. For example, to obtain the GPA boxplots by class year (see Figure 6.5):

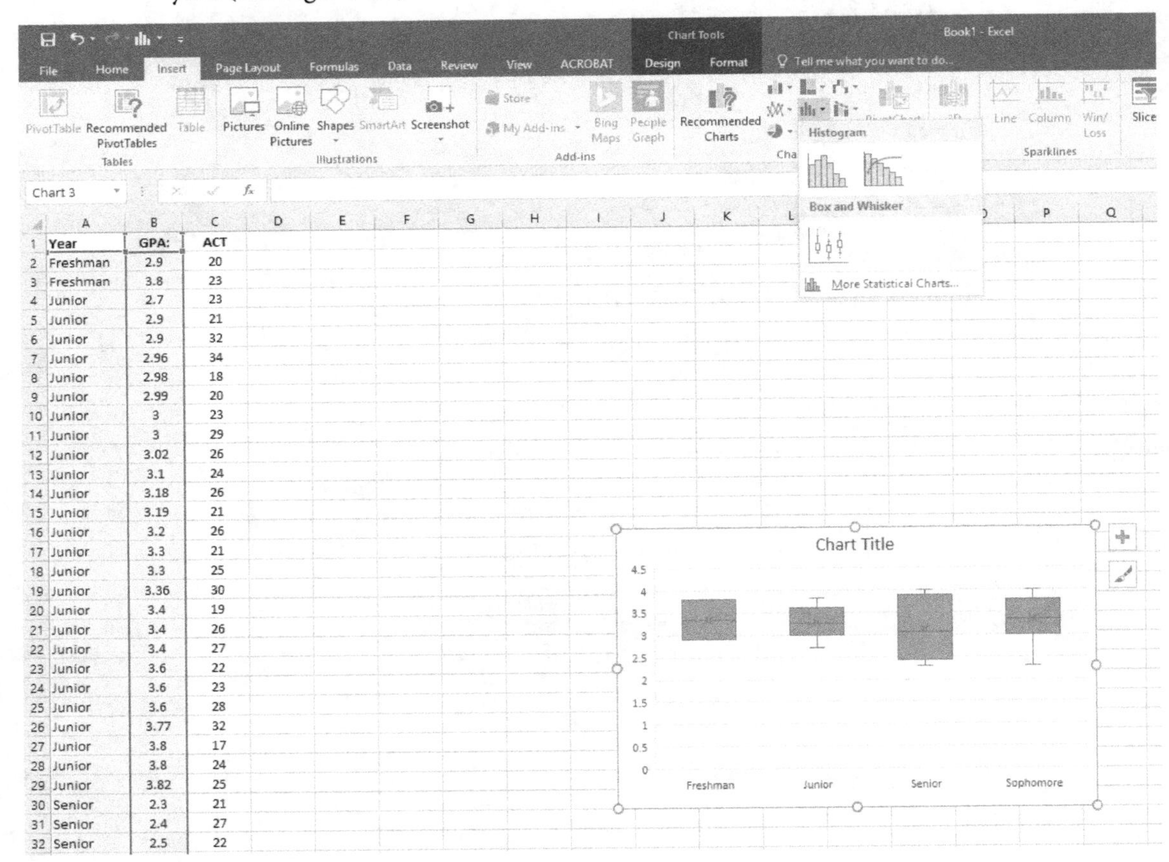

	A	B	C
1	Year	GPA:	ACT
2	Freshman	2.9	20
3	Freshman	3.8	23
4	Junior	2.7	23
5	Junior	2.9	21
6	Junior	2.9	32
7	Junior	2.96	34
8	Junior	2.98	18
9	Junior	2.99	20
10	Junior	3	23
11	Junior	3	29
12	Junior	3.02	26
13	Junior	3.1	24
14	Junior	3.18	26
15	Junior	3.19	21
16	Junior	3.2	26
17	Junior	3.3	21
18	Junior	3.3	25
19	Junior	3.36	30
20	Junior	3.4	19
21	Junior	3.4	26
22	Junior	3.4	27
23	Junior	3.6	22
24	Junior	3.6	23
25	Junior	3.6	28
26	Junior	3.77	32
27	Junior	3.8	17
28	Junior	3.8	24
29	Junior	3.82	25
30	Senior	2.3	21
31	Senior	2.4	27
32	Senior	2.5	22

Figure 6.5.

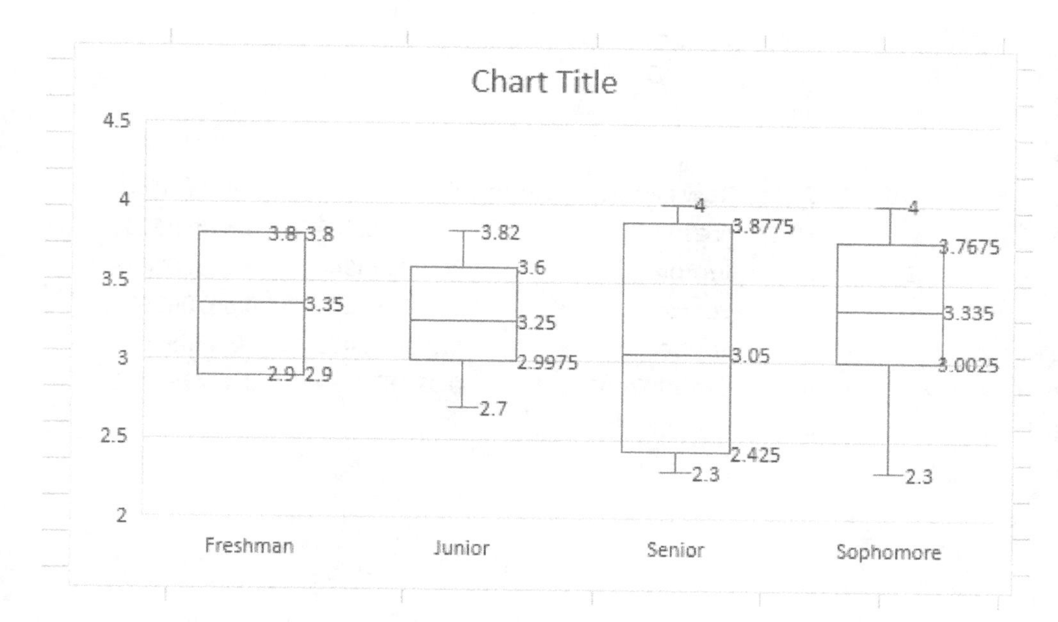

Figure 6.6.

- First, highlight or select the data in columns A (Year) and B (GPA) and then use the boxplot chart to produce the boxplots. Excel will produce as many boxplots as there are different categories in the Year column.
- You can edit the boxplots as we did before by changing the box filling and adding the values of the five-number summary (see Figure 6.6).

MEASURES OF DISPERSION BY CATEGORIES USING PIVOT TABLES

6.6

As with the case for means, pivot tables can compute variance and standard deviations by groups of observations. In our example, we would like to compute the variance and standard deviation for GPA according to class year (see Figure 6.7).

- Click the INSERT tab in the top Excel menu and select the Pivot Table icon.
- In the Create Pivot Table menu, select as range all the values in column A (including the header "Year") and all the values in column B (including the header "GPA"). As the location for the table, click on "Existing worksheet" and enter cell D66 as the location for the upper-left corner of the table. Then press OK.

	A	B	C	D	E	F
64	Sophomore	3.7	29			
65	Sophomore	3.79	31			
66	Sophomore	3.8	27	**Row Labels**	**Var of GPA:**	**StdDev of GPA:**
67	Sophomore	3.8	30	Freshman	0.405	0.636396103
68	Sophomore	3.82	29	Junior	0.104523846	0.323301479
69	Sophomore	3.9	26	Senior	0.485898214	0.697063996
70	Sophomore	4	28	Sophomore	0.208571429	0.45669621
71	Sophomore	4	31	**Grand Total**	**0.197830282**	**0.444781162**
72	Sophomore	4	33			
73	Sophomore	4	34			
74						

Figure 6.7.

- On the Pivot Table Fields menu, drag "Year" (i.e., the header of column A) to the ROWS box. Then, drag "GPA" (the header of column B) to the VALUES box. Click on the down arrow point next to the "Sum of" label and open the "Value Field Settings." In the menu, select "Var" and then OK. Drag again "GPA" to the "VALUES" box. Now, change the value to "StdDevt."

6.7 WORKED EXAMPLE

In the second worksheet of the Excel file for this chapter, we have again the data for class, GPA, and ACT. Now we will analyze the ACT data for the students (see Figure 6.8).

1. Obtain and report the mean, standard deviation, and count for ACT by class year (use a pivot table). Which year has the largest standard deviation? What is the value of this standard deviation? Which year has the smallest standard deviation? What is the value of this standard deviation?
   - Click on the Pivot Table icon. Select as range all the three columns (including the headers). For the location of the table, click "Existing worksheet" and enter the table in the upper-left corner of cell E1.

	A	B	C
1	Year	GPA:	ACT
2	Freshman	2.9	20
3	Freshman	3.8	23
4	Junior	2.7	23
5	Junior	2.9	21
6	Junior	2.9	32
7	Junior	2.96	34
8	Junior	2.98	18
9	Junior	2.99	20
10	Junior	3	23
11	Junior	3	29
12	Junior	3.02	26
13	Junior	3.1	24
14	Junior	3.18	26
15	Junior	3.19	21
16	Junior	3.2	26
17	Junior	3.3	21
18	Junior	3.3	25
19	Junior	3.36	30
20	Junior	3.4	19
21	Junior	3.4	26
22	Junior	3.4	27
23	Junior	3.6	22
24	Junior	3.6	23
25	Junior	3.6	28
26	Junior	3.77	32
27	Junior	3.8	17
28	Junior	3.8	24
29	Junior	3.82	25
30	Senior	2.3	21
31	Senior	2.4	27
32	Senior	2.5	22
33	Senior	3	23

Row Labels	Average of ACT	StdDev of ACT	Count of ACT
Freshman	21.5	2.121320344	2
Junior	24.69230769	4.379673328	26
Senior	25.25	3.150963571	8
Sophomore	26.02777778	3.938717058	36
Grand Total	**25.33333333**	**4.031565593**	**72**

Min	17
Q1	23
Median	25.5
Q3	28
Max	34
IQR	5

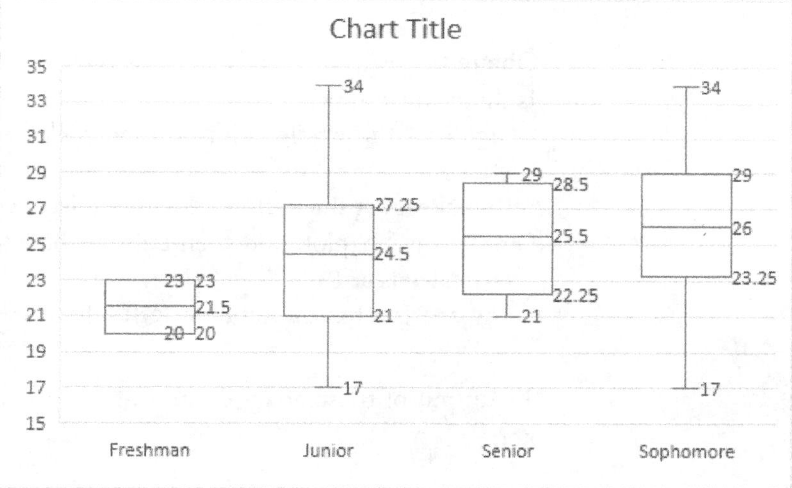

Chart Title

35
33
31
29
27
25
23
21
19
17
15

Freshman: 23, 23, 21.5, 20, 20
Junior: 34, 27.25, 24.5, 21, 17
Senior: 29, 28.5, 25.5, 22.25, 21
Sophomore: 34, 29, 26, 23.25, 17

Freshman Junior Senior Sophomore

Figure 6.8.

- On the Pivot Table Fields menu, drag "Year" to the ROWS box. Then, drag "ACT" to the VALUES box, and change the value to mean. Drag again "ACT" to the VALUES box and change the value to "StdDevt." Drag again "ACT" to the VALUES box and change the value to count.

Juniors, with a value of 4.38, have the largest standard deviation. Freshman have the smallest standard deviation, 2.12.

2. Use the Excel functions to obtain the five-number summary for ACT. What is the value of the first and third quartile? What is the value of the interquartile range?
   - In cell F8, enter the function =MIN(C2:C73).
   - In cell F9, enter the function =QUARTILE(C2:C73,1).
   - In cell F10, enter the function =MEDIAN(C2:C73).
   - In cell F11, enter the function =QUARTILE(C2:C73, 3).
   - In cell F12, enter the function = MAX(C2:C73).
   - In cell F14, the IQR is the difference between the third and first quartile, =F11-F9.

   The first quartile is 23 and the third quartile is 28. The IQR is 5.

3. Obtain the boxplot of ACT by class year. Which class year has the smallest IQR? Is there any outlier?
   - Highlight the whole column A by clicking on the gray header cell with the letter A.
   - Then, keeping the Ctrl key down highlight column C.
   - Insert the boxplot, and then edit the boxes as we described before. Start the vertical axis at 15.

   Freshman class has the smallest IQR. There are no outliers.

The output of these analyses are in the third worksheet of the Excel file for this chapter.

7 Measures of Location: Percentiles and Standard Scores

Previous chapters showed procedures to summarize information from numerical observations. We can talk, for example, of the average verbal achievement of children in the fifth grade or the variability in height of male teenagers. However, in many situations we need to focus on a particular individual observation in the data set. For example, we would like to know how well Mary's verbal achievement is in comparison to other fifth graders; or how tall John is with respect to children of his age. Measures of location, such as percentiles and standard scores, help us to answer questions like that.

Measures of location are numbers that indicate the position of a particular value within a data set. For example, a **percentile** expresses the location of a value in terms of the percentage of observations below that value. A ***standard score*** expresses the location of a value in terms of how far the average is from the data in units determined by the standard deviation of the data. Measures of location are widely used to answer questions such as:

- Meghan has a raw score of 45 in a standardized verbal test. Is this a good score?
- Baby Sue weighs at the percentile 55 for 9-month-old babies. Is she developing in a normal fashion?
- Diego needs to be one standard deviation above the average speed of teenagers in the 100 meters. What is the minimum speed that he needs to achieve?

7.1 OBJECTIVES

1. To distinguish between percentile points and percentile ranks and determine how to obtain them using Excel functions.
2. To define standard scores and determine how to interpret and calculate them using Excel functions.

7.2 PERCENTILES

The ***percentile point*** (or simply percentile) ***p%*** is the ***score*** that splits the data in the lower p% and the upper (100 – *p*%) of the observations. Thus, the 32% percentile (point) is the score that divides the distribution of the data in the lower 32% of the values and the upper 68% of the values. The complementary measure is the ***percentile rank.*** This is the ***percentage*** of observations in the distribution that are below or at the same level of a given score value. For example, the percentile rank for a raw score of 19 is the percentage of observations in the data that have a score of 19 or less. The percentile rank is similar to the ***cumulative percent*** in a frequency distribution.

Excel has two functions for requesting the percentile score for a particular percentage p% (or proportion) in a data range: PERCENTILE.EXC and PERCENTILE.INC (see Figure 7.1). There is a legacy function, PERCENTILE, that produces the

	A	B	C	D	E	F
1	DATA			p	PERCENTILE.EXC	PERCENTILE.INC
2	1		100%	1	#NUM!	10
3	2		N/(N+1)	0.909	9.999	9.181
4	3		Q3	0.75	8.25	7.75
5	4		Median	0.5	5.5	5.5
6	5		Q1	0.25	2.75	3.25
7	6		1/(N+1)	0.091	1.001	1.819
8	7		0%	0	#NUM!	1
9	8					
10	9					
11	10					

Figure 7.1.

same output as the function PERCENTILE.INC. The PERCENTILE.EXC function does not compute a percentile score when the percentage or p% is below $1/(N + 1)$ or above $N/(N + 1)$; in other words it will not provide a percentile score for 100% or 0%. On the other hand, the PERCENTILE.INC will return the maximum score in the data set as the 100 percentile point and the lowest score in the data set as the 0 percentile point. The first worksheet of the Excel file for this chapter shows the different behavior of the two functions in a simple data set of $N = 10$ observations. Thus, the 100% percentile point for the data is the maximum score of 10 when using PERCENTILE.INC, but the same value is not defined when using PERCENTILE.EXC.

The Excel functions PERCENTRANK.EXC and PERCENTRANK.INC (which work in the same way as the legacy function PERCENTRANK) compute the percentile rank or percentage for a score in the data range. When we have repeated values in a data set, the percentile rank functions compute the percentile rank by using the first occurrence of the repeated value. This may produce that the percentile rank of the maximum value in the data does not reach 100%.

In the second worksheet in the Excel file for this chapter, we have the American College Test (ACT) scores of 135 college freshman. The scores are in column A with the header "ACT." Suppose that we would like to create a table that attaches to each ACT score in the range of the data the corresponding percentile rank (see Figure 7.2).

- First, to ease the computations, we name the data in column A. Highlight the column of values, right click on the mouse, select the "Define Name" option, and then enter the name ACT.
- To obtain the range of values in the data, we find the minimum and maximum scores in the data set. In our case =MIN(ACT) is 14 and =MAX(ACT) is 36. Thus, our table should have entries from 14 to 36.
- We create a column of sequential scores from 14 to 36 starting in the cell E2. We enter 14 and 15 as the first two entries, and then drag the two cells to fill the rest of cells up to the value 36. We label the column ACT score.
- Now we put in column F2 the percentile rank (or percentage) that corresponds to the ACT score of 14 in E2.

  = PERCENTRANK.EXC(ACT, E2)

	A	B	C	D	E	F
1	ACT				ACT Score	PR
2	31	Min	14		14	0.70%
3	33	Max	36		15	1.40%
4	32				16	2.90%
5	35				17	3.30%
6	35				18	3.60%
7	27				19	3.90%
8	34				20	4.10%
9	36				21	4.40%
10	28				22	5.80%
11	16				23	6.60%
12	27				24	9.50%
13	34				25	12.50%
14	28				26	14.70%
15	33				27	17.60%
16	32				28	25.00%
17	34				29	36.70%
18	28				30	45.50%
19	35				31	56.60%
20	33				32	64.70%
21	30				33	76.40%
22	31				34	86.00%
23	34				35	92.60%
24	32				36	98.50%
25	30					
26	28					

Figure 7.2.

Notice that the range name ACT is actually an absolute reference. Drag this formula to the remaining cells to complete the table. The results will be proportions. To convert the column to percentages, select the column, click the mouse right button, select "Format Cells," and select "Percentage." Add as a header to column PR.

Thus, a score of 25 in the ACT, puts the student in the percentile rank of 12.5% of the data set. Meanwhile, a score of 30 corresponds to a percentile rank of 45.50%. The percentile rank formula is actually an interpolation using cumulative frequencies. In order to avoid going above 100%, the formula truncates the percentile rank of the maximum scores. Thus, in the table the percentile rank for 100 is only 98.5%.

	A	B	C	D	E	F	G	H	I
1	ACT				ACT Score	PR		%P	Score
2	31	Min	14		14	0.70%		0.95	35
3	33	Max	36		15	1.40%		0.9	34
4	32				16	2.90%		0.85	33
5	35				17	3.30%		0.8	33
6	35				18	3.60%		0.75	32
7	27				19	3.90%		0.7	32
8	34				20	4.10%		0.65	32
9	36				21	4.40%		0.6	31
10	28				22	5.80%		0.55	30
11	16				23	6.60%		0.5	30
12	27				24	9.50%		0.45	29.2
13	34				25	12.50%		0.4	29
14	28				26	14.70%		0.35	28
15	33				27	17.60%		0.3	28
16	32				28	25.00%		0.25	28
17	34				29	36.70%		0.2	27
18	28				30	45.50%		0.15	26
19	35				31	56.60%		0.1	24
20	33				32	64.70%		0.05	21
21	30				33	76.40%			
22	31				34	86.00%			
23	34				35	92.60%			
24	32				36	98.50%			
25	30								

Figure 7.3.

Suppose that we want to now build a table with the percentile points or scores for selected values of percentile ranks or percentages. For example, we want to build a table for the percentiles in 5% steps, from the 5% percentile to the 95% percentile (see Figure 7.3).

- First, we create a column of sequential p% proportions from 0.95 to 0.05 in column H. Again, the easy way is to enter 0.95 and 0.90 as the first two entries, and then drag the two cells to fill the rest of cell up to the value 0.05.
- Then, we create a new column of percentiles scores in I. For cell I2 that corresponds to a p% of 0.95 or 95% point we enter: = PERCENTILE.EXC(ACT, I2) and drag this formula to the next cells to complete the table.

Thus, to be in the 95% percentile, a student needs to have a score of 35 or more. In similar manner, to be in the 50% percentile, a student needs a score of 30. The table also tells that 32 is the 75th percentile, or the third quartile. Similarly, 28 is the 25th percentile of first quartile.

7.3 STANDARD SCORES Z

The standard score, *Z*, expresses the location of an individual value in a data set as a function of how far the value is from the mean of the data. This distance, or deviation, is expressed in units of the standard deviation of the data.

For example, a group of female teenagers has an average weight of 115 pounds with a standard deviation of 5 pounds. Mary's weight is 110 pounds. Thus, Mary is 5 pounds *below* the average. However, 5 pounds is the actual standard deviation of the data, thus Mary is one standard deviation below the average, or a *Z* score of −1. Rachel's weight is 120 pounds. She is 5 pounds above the *average*, and because the standard deviation is 5, Rachel is one standard deviation above the average, or a *Z* score of +1. Margaret's weight is 115 pounds and therefore she is not above or below the mean; she is exactly at the mean value. Her *Z* score will be in this case 0.

The formula to obtain the standard score of an individual raw score X is given by:

$$Z_i = \frac{X_i - \bar{X}}{S}$$

In Excel, the function STANDARDIZE (value, mean, standard deviation) computes the *Z* score once you provide as input a raw score value and the mean and standard deviation of the data.

To produce the standard scores for each ACT score from 14 to 36, we will do the following:

In the third worksheet of the Excel file for this chapter, we obtain the standard scores for the raw scores that we found in the ACT data (see Figure 7.4). In the first worksheet, we found that the ACT values range from 14 to 36.

- Enter the values from 14 to 36 in column D.
- To obtain the *Z* scores for these values, we need first to compute the mean and standard deviation of the original ACT data. We enter in cells B2 and B3, respectively, the functions.

  =AVERAGE(ACT)
  =STDEV.S(ACT)

	A	B	C	D	E	F	G
1				ACT Score	Z score	T scores	IQ scores
2	Mean	29.37037		14	-3.57897	14	43
3	S.D	4.294635		15	-3.34612	17	46
4				16	-3.11327	19	50
5				17	-2.88042	21	54
6				18	-2.64758	24	58
7				19	-2.41473	26	61
8				20	-2.18188	28	65
9				21	-1.94903	31	69
10				22	-1.71618	33	73
11				23	-1.48333	35	76
12				24	-1.25048	37	80
13				25	-1.01763	40	84
14				26	-0.78479	42	87
15				27	-0.55194	44	91
16				28	-0.31909	47	95
17				29	-0.08624	49	99
18				30	0.146608	51	102
19				31	0.379457	54	106
20				32	0.612306	56	110
21				33	0.845154	58	114
22				34	1.078003	61	117
23				35	1.310852	63	121
24				36	1.5437	65	125
25							

Figure 7.4.

Notice that we call the ACT named range from the second worksheet. Thus, named ranges are available in any worksheet in the same file.

- To compute the Z score for the value 14 in cell D2, we enter in cell E2 the function:

  =STANDARDIZE(D2, $B$2, $B$3)

Notice that the cells referring to the mean (B2) and standard deviation (B3) have an absolute reference.

- Drag the function to the remaining cells in column E.

Thus, a student with the lowest score of 14 has a standard score −3.57897, i.e., the student is **below** the average of the data by over three and half standard deviations. On the other hand, a student with a score of 36 has a standard score of 1.54, i.e., the student is **above** the average of the data by over one and half standard deviations.

7.4 OTHER STANDARD SCORES

Z scores can be positive or negative values, with 0 representing the mean and with the standard deviation being represented by 1. If John has *Z* score of −0.5, we know that *"John is below the mean by half standard deviation."* Can we convey the same type of information but avoid numbers that are negative and decimal? Yes. *Z* scores are not the only type of standard scores. We have, among others, the *T* scores and IQ scores.

T scores represent the mean of the data with a value of 50 and the standard deviation with a value of 10.

IQ scores represent the mean of the data with a value of 100 and the standard deviation with a value of 16.

Thus, *"being half standard deviation below the mean"* will be a *Z* score of −0.5, or a T score of 45, or an IQ score of 92.

Once we have the *Z* scores, we can obtain the other standardized scores very easily (see Figure 7.4).

- To obtain the ***T scores*** (mean 50 and standard deviation 10), we enter in cell F2 the expression =50+10*E2, and drag it to the remaining cells. We usually round *T* scores to integers. (Select the values in column F and use the "Number" menu at the top bar in the Home tag to reduce the number of decimals.)
- To obtain the ***IQ scores*** (that have mean 100 and standard deviation 16), we enter in cell G2 the expression = 100 + 16*E2 and drag it to the remaining cells. Again, we express the results in integers.

Thus, the raw score of 29, with a *Z* score of −.08624 (close to the mean 0 for the *Z* scores), has a *T* score of 49 and IQ score of 99.

WORKED EXAMPLE

7.5

In the fourth worksheet of the Excel file for this chapter, we have 100 new ACT scores in column A, but now sorted from the smallest to the largest.

1. We want to obtain the percentile points (scores) that correspond to the 95%, 90%, 75%, 50%, 25%, 10%, and 5% percentile ranks (see Figure 7.5).

To be in the 95% percentile rank of the distribution, what is the score that a student has to have?

	A	B	C	D	E	F	G	H
1	NewScore		PR	Score		Scores	Z	T
2	16		0.95	34		16	-3.132518801	19
3	18		0.9	31		17	-2.851323217	21
4	20		0.75	29		18	-2.570127634	24
5	21		0.5	27		19	-2.28893205	27
6	21		0.25	25		20	-2.007736466	30
7	22		0.1	23		21	-1.726540883	33
8	22		0.05	21.05		22	-1.445345299	36
9	22					23	-1.164149716	38
10	22					24	-0.882954132	41
11	23		For 30->	0.787		25	-0.601758549	44
12	23					26	-0.320562965	47
13	23		Min	16		27	-0.039367382	50
14	23		Max	36		28	0.241828202	52
15	24					29	0.523023785	55
16	24					30	0.804219369	58
17	24		Mean	27.14		31	1.085414952	61
18	24		SD	3.556244		32	1.366610536	64
19	24					33	1.64780612	66
20	24					34	1.929001703	69
21	24					35	2.210197287	72
22	24					36	2.49139287	75
23	24							

Figure 7.5.

To be in the 10% percentile rank, what is the score that a student has to have?

- We name the range of values in A as "newscore." We enter the values 0.95, 0.90, etc., in column C, then in cell D2, we enter =PERCENTILE.EXC(newscore, C2) and drag the function to the remaining cells

To be in the 95% percentile, a student needs a score of 34. To be in the 10% percentile rank, a student needs a score of 23.

2. A student with a new score of 30 wants to know her/his percentile rank. Using the PERCENTRANK.EXC function finds the estimated percentile rank for this student.
    - Enter in cell D11 the function =PERCENTERANK(newscore, 30)

A percentile rank of 0.787 or 78.7%

3. Find the minimum and maximum ACT values in the data and create a column of all the values between those two values. Put the values in column F (start the first value in F2). Compute the mean and standard deviations of the data, and put in column G, the corresponding Z score.

To be below the mean by two standard deviations or more, what is the raw ACT score that a student needs to have?

To be above the mean by one standard deviation or more, what is the raw ACT score that a student needs to have?

- Find the minimum and maximum of the newdata in cells D13 and D14, respectively.
- Starting in F2, enter the values from the minimum to the maximum.
- Find the mean and standard deviation of the newscore data in D17 and D18, respectively.
- In G2, enter the function =STANDARDIZE(F2, $D$17, $D$18) and drag the function to the remaining cells. Label the column Z.

To be below the mean by two standard deviations or more, the ACT raw score is 20 or less.

To be above the mean by one standard deviation or more, the ACT raw score is 31 or more.

4. Add in column H the *T* scores, rounded to integers. What will be the *T* score for a student with ACT score of 25? What will be the raw ACT score for a student with a *T* score of 69?
   - In cell H2, enter =round(50+10*G2, 0) and drag the function to the remaining cells. Label the column T.

   T score of 44 corresponds to a raw ACT score of 25.
   Raw ACT score of 34 corresponds to a *T* score of 69.

The results of these analyses are in the fifth worksheet of the Excel file for this chapter.

8 Normal Distribution

In previous chapters, we introduced procedures for describing and summarizing data. For example, how to obtain the average number of daily hours that a group of college students works in a paid job, or the histogram for the grade point average (GPA) of a group of college students. In all these procedures, the actual distribution of the values is the main source of information, because, knowing the distribution, we can compute mean, standard deviation, percentiles, histograms, etc. However, what if we want to go beyond our data; e.g., if we want to know the average number of hours that college students work in the nation, or the histogram of the GPA of all college seniors in a state. These questions focus on the generalization of our data to wider settings and introduce us to concepts such as *population*, *sample*, *descriptive*, and *inferential* statistics, as well as the important concept of *normal distribution*.

8.1 OBJECTIVES

1. Assuming a normal distribution, use Excel functions to find the percentile rank of an individual Z score.
2. Assuming a normal distribution, use Excel functions to find the Z score of an individual percentile rank.
3. Check assumption of normality of data using a PP plot or normal probability plot.

8.2 SAMPLE, POPULATION, AND DESCRIPTIVE AND INFERENTIAL STATISTICS

A basic mental schema for data analysis can be summarized as follows: "Use sample data from an unknown population and by analyzing the data acquire knowledge about that population." In statistics and data analysis, ***population*** refers to the set of all individuals, objects, or events from which we are interested in knowing something. For example, we might determine paid job hours among all college students in the nation or distribution of GPA for all college seniors in the nation. In many circumstances, it is impossible to obtain data from the whole population. However, we may obtain data from a ***sample***, i.e., a subset of elements from our desired population. We will see later that a good sample always involves a random selection component.

Using data from a sample, we compute different ***descriptive statistics*** such as the mean, standard deviation, percentiles, histograms, etc. We call these indicators ***statistics*** to emphasize that their values may differ if we use another sample from the same population. If for some magical procedure, we are able to know the mean, standard deviation, percentiles, histograms, etc. of the population, we will call these indicators ***parameters***, to stress that they are values that are not subject to sampling variability. Thus, we talk of ***sample statistics*** and ***population parameters***. We use the sample data to *infer* characteristics for the population. ***Inferential statistics*** comprise the procedures for inferring values of population parameters using as input sample statistics.

8.3 NORMAL DISTRIBUTION AS A MODEL FOR POPULATION DISTRIBUTION

The distribution of the values in a sample is the main source of information for analysis. In a sample, we describe the distribution of values using a frequency distribution or a histogram. However, do we know the shape of the distribution of values in the population from which we get our sample? Well, in many situations, we actually do not know the shape of the distribution. However, we make assumptions about this shape, and use our data to check if the assumption is justifiable. We use mathematical distributions to model the possible population distributions. These mathematical functions are called ***probability distributions***.

There are many probability distributions, but we will focus in this chapter on only one of them, the ***normal distribution***. The normal distribution is a probability distribution for a continuous variable where the majority of values concentrates around the mean and smaller proportion of values locates on the "tails" of the distribution, i.e., values well below the mean or values well above the mean. These properties are reflected in the "bell-curve" shape of the distribution (see Figure 8.1). The shape and location of the normal distribution is a function of its mean and standard deviation. Thus, if we express the data in standard scores, the mean of the ***standard normal distribution*** is zero, with standard deviation of one. Continuous probability distributions are represented by mathematical functions that relate the value of a variable with its ***likelihood*** (sometimes called ***density***). The actual probabilities for a range of values is defined as the **area** under the distribution. We can interpret these areas as probabilities or as proportion or percentage of cases that take values in that range.

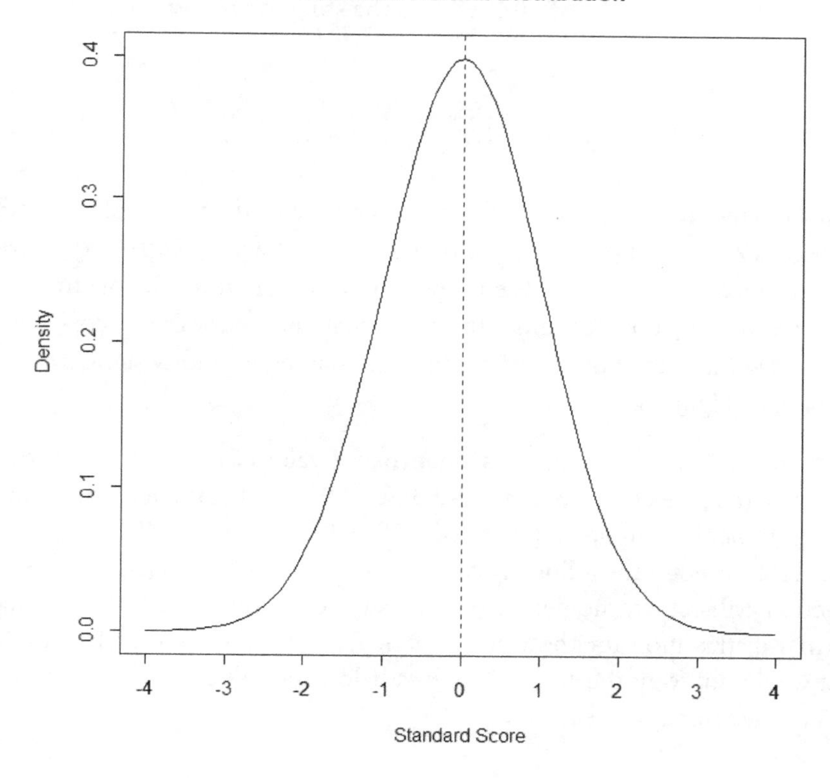

Figure 8.1.

Many empirical data distributions (i.e., people's heights, scores in well-designed ability tests, biometric dimensions, etc.) approximate the normal distribution shape. We will see later, that the assumption of normality, i.e., of a normal distribution of the values, will be essential for inferential procedures. In this chapter, we will focus on how a normal distribution helps to link standard scores and percentile scores and ranks. For example, if the data follow a normal distribution, nearly all the values (actually 99.73%) will be between three standard deviations below or above the mean, i.e., between standard scores of −3 and +3. Also, if the data follow a normal distribution, an individual with a standard score of $Z = +1$, will be in the 84.13% percentile rank. Also, in order to be in the top ten percent of the distribution, i.e., in percentile 90%, an individual needs to have a standard score of at least $Z = +1.281552$. Thus, if we know that a data set follows a normal distribution, we can easily answer, among other things, questions such as:

- If the percentile rank of an individual is 45%, what is the person's Z score?
- If a student has a Z score of −0.5, what is the student's percentile rank?

8.4 STANDARD NORMAL DISTRIBUTION

A ***standard normal distribution*** is a function that relates the values of a variable in standard scores Z (on the horizontal axis of the function) with a ***density*** or ***likelihood*** (or the vertical axis of function). The proportion of observations below the Z score value is represented by the area under the curve that falls below the Z score.

In the first worksheet of the Excel file for this chapter, we plot a standard normal distribution (see Figure 8.2).

- First, starting in A2, we enter a sequence of Z values from −3 to +3 in steps of 0.01. For this, we enter −3 in cell A2 and −2.9 in cell A3. Highlight both cells and drag them down up to a value of 3. We label the column Z.
- In cell B2, we enter the following command: =NORM.S.DIST(A2, FALSE). This function will return the height (or "density" or "likelihood") of the standard normal distribution for the value of Z in A2. The statement FALSE indicates that we do not want the cumulative area below the value in A2. Drag the function to the remaining cells in B.

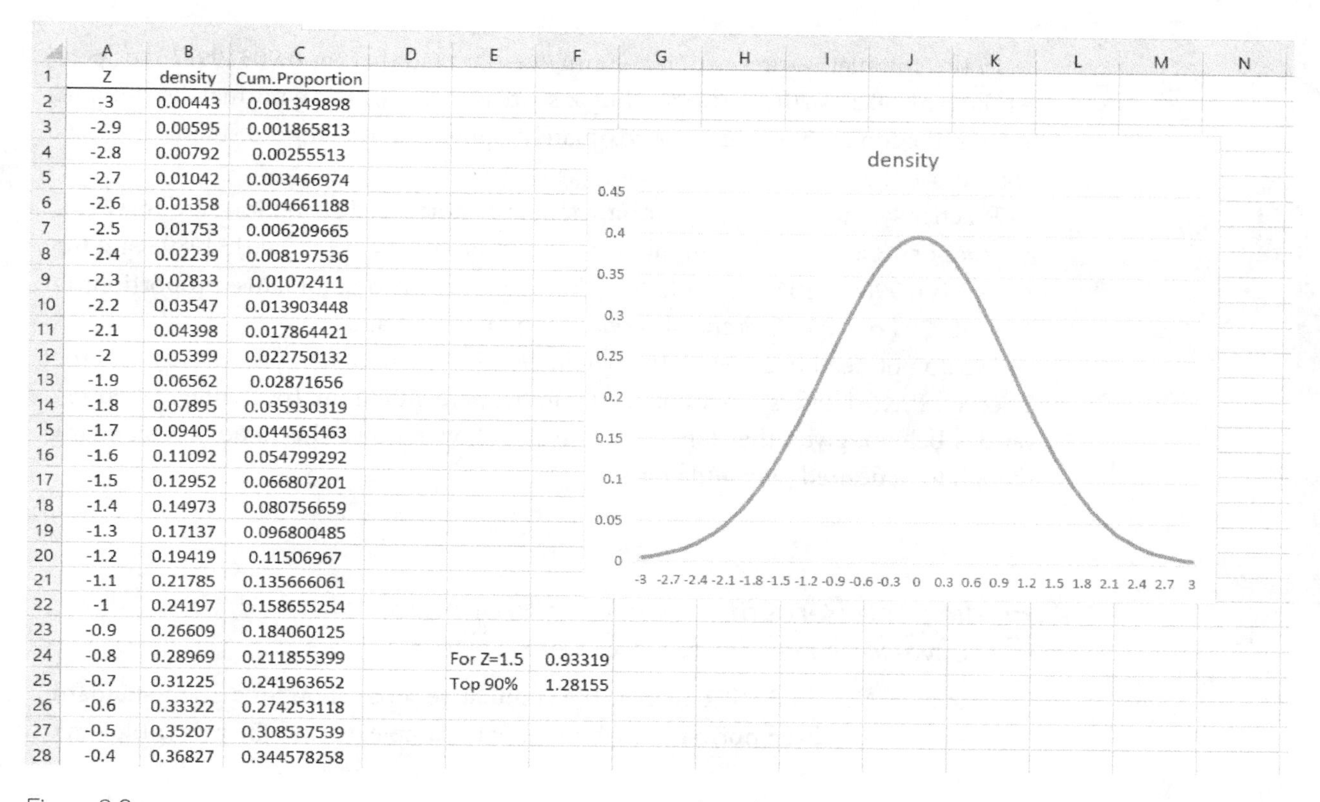

	A	B	C	D	E	F
1	Z	density	Cum.Proportion			
2	-3	0.00443	0.001349898			
3	-2.9	0.00595	0.001865813			
4	-2.8	0.00792	0.00255513			
5	-2.7	0.01042	0.003466974			
6	-2.6	0.01358	0.004661188			
7	-2.5	0.01753	0.006209665			
8	-2.4	0.02239	0.008197536			
9	-2.3	0.02833	0.01072411			
10	-2.2	0.03547	0.013903448			
11	-2.1	0.04398	0.017864421			
12	-2	0.05399	0.022750132			
13	-1.9	0.06562	0.02871656			
14	-1.8	0.07895	0.035930319			
15	-1.7	0.09405	0.044565463			
16	-1.6	0.11092	0.054799292			
17	-1.5	0.12952	0.066807201			
18	-1.4	0.14973	0.080756659			
19	-1.3	0.17137	0.096800485			
20	-1.2	0.19419	0.11506967			
21	-1.1	0.21785	0.135666061			
22	-1	0.24197	0.158655254			
23	-0.9	0.26609	0.184060125			
24	-0.8	0.28969	0.211855399		For Z=1.5	0.93319
25	-0.7	0.31225	0.241963652		Top 90%	1.28155
26	-0.6	0.33322	0.274253118			
27	-0.5	0.35207	0.308537539			
28	-0.4	0.36827	0.344578258			

Figure 8.2.

- In cell C3, we enter =NORM.S.DIST(A2, TRUE). Now the function will return the cumulative proportion (or "area" under the curve) below the standard score value of Z in A2. Drag the function to the remaining cells in C.
- Finally, we highlight both columns A and B and, click the Insert tag in the top Excel menu, and then in the Chart menu we click on "Recommended Charts" and pick up the one that uses lines (the second in the "Recommended Charts" list.

The total area under the standard normal distribution represents a proportion of 1, or the 100%, of the values. Thus, the areas under the standard normal relates to the standard scores, Z, with their corresponding percentile ranks.

For example, for the standard score of $Z = -1$, we have a cumulative proportion of 0.158655254 or percentile rank of 15.87% approximately. Also, for a Z score of 0, i.e., for the mean, the cumulative proportion is 0.5 or a percentile rank of 50%.

In the cumulative proportion column, we can read that for 0.903199515, or percentile rank 90.32%, the corresponding *Z* score is +1.3. Thus, in a normal distribution to be in the top ten percent is approximately equivalent to being 1.3 standard deviations above the mean.

Practically, all the values in a standard normal distribution are between *Z* scores of −3 and +3. The cumulative proportion for $Z = -3$ is 0.001349898 and for $Z = +3$ is 0.998650102. The difference between these two cumulative proportions is 0.997300204 or 99.73% of the observations between −3 and +3.

We do not need to generate the whole distribution to obtain percentile ranks for *Z* scores. EXCEL offers functions to obtain the percentile rank (percentage of observations below), given the standard score *Z* and, vice versa, the value of a standard score *Z* for a desired percentile rank.

Finding the Percentile Rank (or Proportion of Observations Below) for a Given Z Score

The function NORM.S.DIST returns the cumulative proportion for a given standard score in a standardized normal distribution. For example, to find the percentile rank for a *Z* score of +1.5:

- Enter in cell F24 =NORM.S.DIST(1.5, TRUE). The parameter TRUE means that we want cumulative proportions. Excel returns the value 0.933193, or approximately a percentile rank of 93.32%.

The function NORM.DIST is a more general function to obtain cumulative proportions that requires us to provide the mean and standard deviation of the distribution. For example, =NORM.DIST(1.5, 2, 0.5, TRUE) requests the cumulative proportion for a value 1.5 in a normal distribution with mean 2 and standard deviation 0.5. The result is 0.15866 or 15.87%.

Finding the Z Score for a Given Percentile Rank

The function NORM.S.INV returns the *Z* score for a given cumulative proportion in a standardized normal distribution. For example, to find the *Z* score that defines the top 10% of the values on a standard normal distribution (i.e., the 90% percentile or cumulative proportion 0.9):

- Enter in cell F25 =NORM.S.INV(.90) and Excel returns a *Z* score of 1.28155.

The function NORM.INV is a more general function to obtain the score that requires us to provide the mean and standard deviation of the distribution. For example, = NORM.INV(0.9, 100, 16) requests the score that defines the 0.9 cumulative proportion, or 90%, in a normal distribution with mean 100 and standard deviation 16. The result is 120.505.

8.5 CHECKING IF THE DATA FOLLOWS A NORMAL DISTRIBUTION SHAPE

To use the normal distribution as our model for linking standard scores and percentile ranks, we need the data to be approximately normal. We can check if the empirical data in a sample follows approximately a normal distribution by comparing the actual percentile rank of the sample values against the expected percentile ranks provided by a normal distribution.

For example, in the second worksheet of the Excel file for this chapter, we have the height of 57 students in a class and the cash that the students have in their pockets (see Figure 8.3). The mean and standard deviation of height are 65.25 and 3.38 (in cells D3 and D4).

We create two columns of percentile ranks:

Column F contains the actual percentile rank ("Actual PR") for each value of height in the data:

- In cell F2, we enter the function PERCENTRANK.EXC($A$2:$5$8, A2) and then drag this formula to the rest of the rows. Notice that range A2:A58 is the raw data in absolute reference.

Column G contains the predicted percentile rank if we assume that the height data follows a normal distribution ("If Normal") with mean 65.25 and standard deviation 3.38.

- In cell G2, we enter the function =NORM.DIST(A2, $D$3, $D$4, TRUE) and then drag this function to the rest of the columns. Notice that we use the mean, in D3, and standard deviation, in D4, and enter them as absolute references.

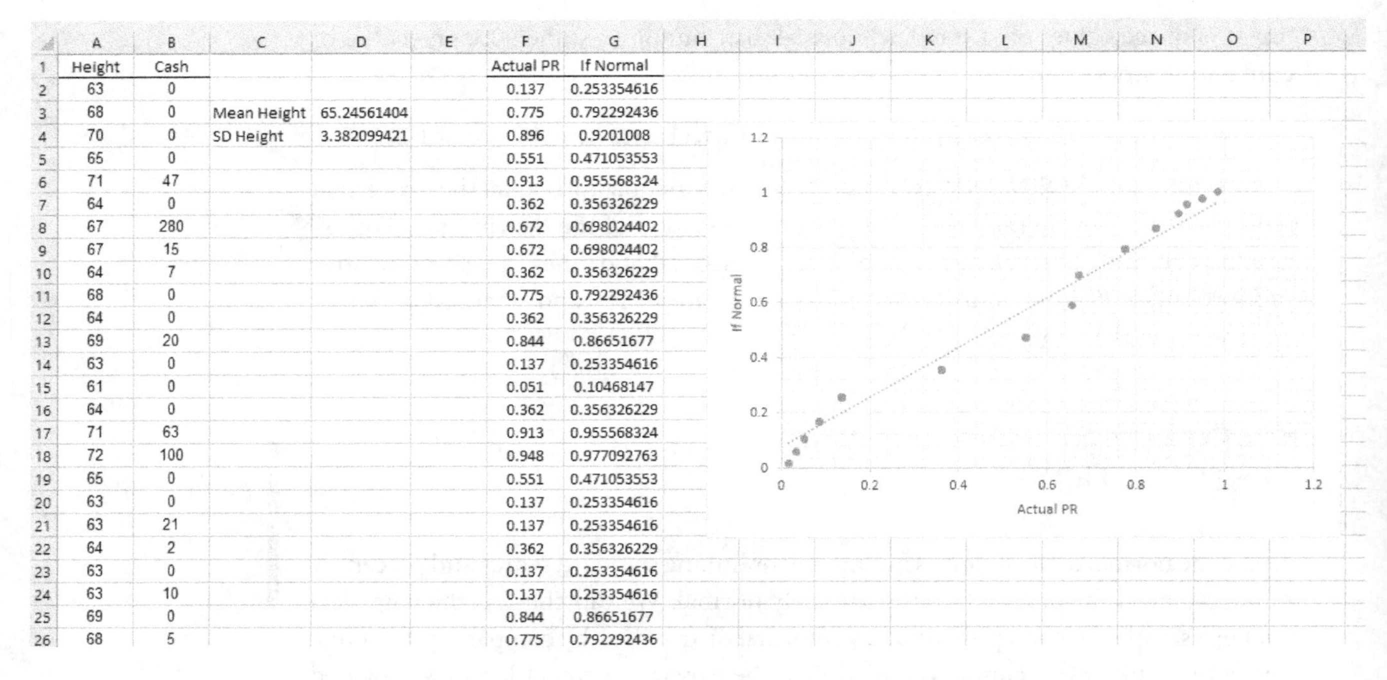

	A	B	C	D	E	F	G
1	Height	Cash				Actual PR	If Normal
2	63	0				0.137	0.253354616
3	68	0	Mean Height	65.24561404		0.775	0.792292436
4	70	0	SD Height	3.382099421		0.896	0.9201008
5	65	0				0.551	0.471053553
6	71	47				0.913	0.955568324
7	64	0				0.362	0.356326229
8	67	280				0.672	0.698024402
9	67	15				0.672	0.698024402
10	64	7				0.362	0.356326229
11	68	0				0.775	0.792292436
12	64	0				0.362	0.356326229
13	69	20				0.844	0.86651677
14	63	0				0.137	0.253354616
15	61	0				0.051	0.10468147
16	64	0				0.362	0.356326229
17	71	63				0.913	0.955568324
18	72	100				0.948	0.977092763
19	65	0				0.551	0.471053553
20	63	0				0.137	0.253354616
21	63	21				0.137	0.253354616
22	64	2				0.362	0.356326229
23	63	0				0.137	0.253354616
24	63	10				0.137	0.253354616
25	69	0				0.844	0.86651677
26	68	5				0.775	0.792292436

Figure 8.3.

To compare the values of the two columns, we use a plot.

- Highlight the "Actual PR" and "If Normal" columns.
- Select the Insert tab in the top Excel menu, check the Recommended Chart, select the Scatter one, and check OK.
- In the Chart Tools tab in the top Excel menu, select the "Design" subtab. In the icon for "Quick Layout" select "Layout 3". In the x-axis type in "Actual PR" and in the y-axis type in "If Normal." Erase the Legends "If Normal" and "Linear(If Normal)."

If the sample data is approximately normal, the values in the two columns should be similar, and the plot should resemble a straight line. This is the case for height. (Note that the x axis of the plot is "Actual PR" and the y axis is the "If Normal.") This type of plot is referred as a ***PP plot*** (percentile vs. percentile) or a ***normal probability plot.***

WORKED EXAMPLE

8.6

We will use now the Cash-in-pocket information. In the third worksheet of the Excel file for this chapter, we have again the height and cash data (see Figure 8.4).

1. Find and report the mean and standard deviation for the cash-in-pocket data.
   - In cell D4, we enter =AVERAGE(B2:B58) and in D5 =STDEV.S(B2:B58)

   The mean is 26 and SD is 80.29617.

2. If we assume that the data follow a normal distribution, what will be the percentile rank for a student with 10 dollars in his/her pocket? What is the actual percentile rank for a student with 10 dollars in his/her pocket? Are these percentages closer to each other?
   - For the percentile rank assuming normality, in cell D8, we enter =NORM.DIST(10, D4, D5, TRUE). The result is 0.42225 or 42.225%
   - For the actual percentile rank of 10 in the data, in cell D9, we enter =PERCENTRANK.EXC(B2:B58, 10). The result is 0.655 or 65.5%.

   No, these two percentiles are not close to each other.

	A	B	C	D	E	F	G
1	Height	Cash				Actual PR	If Normal
2	63	0				0.017	0.374202802
3	68	0	For Cash			0.017	0.374202802
4	70	0	Mean	26		0.017	0.374202802
5	65	0	SD	80.29617		0.017	0.374202802
6	71	47				0.896	0.604337584
7	64	0	For $ 10			0.017	0.374202802
8	67	280	Assuming Norma	0.422225		0.965	0.99922817
9	67	15	Actual PR	0.655		0.689	0.446727386
10	64	7				0.586	0.407661294
11	68	0	For 50th PR			0.017	0.374202802
12	64	0	Assuming Norma	25.75439		0.017	0.374202802
13	69	20	Actual Perc.	2		0.706	0.471434447
14	63	0				0.017	0.374202802
15	61	0				0.017	0.374202802
16	64	0				0.017	0.374202802
17	71	63				0.931	0.67862345
18	72	100				0.948	0.822425237
19	65	0				0.017	0.374202802
20	63	0				0.017	0.374202802
21	63	21				0.81	0.476392175
22	64	2				0.5	0.383678191
23	63	0				0.017	0.374202802
24	63	10				0.655	0.422225458

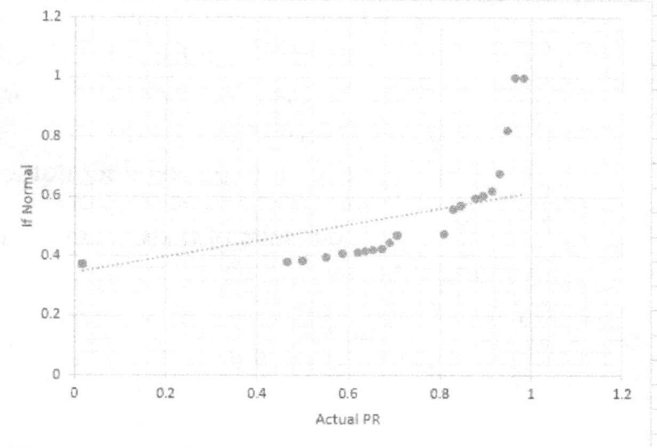

Figure 8.4.

3. Suppose that the data follows a normal distribution: what will be the amount of cash that a student in the 50th percentile rank would carry? In other words, the 50th percentile or percentile point. Find the actual 50th percentile or percentile point in the data. Are these two amounts close to each other?
   - For the 50th percentile assuming normality, in cell D12, we enter =NORM.INV(0.5, D4, D5). The result is approximately 25.75 dollars.
   - For the actual 50th percentile in the data, in cell D13, we enter =PERCENTILE.EXC(B2:B58, 0.5). The result is 2 dollars.

   No, these two percentile points are not close to each other.

4. We want to check if the cash data follows a normal distribution. We obtain a column with the actual percentile ranks ("Actual PR") for each score and a column with the percentile ranks of the values if we assume normality ("If Normal"). Plot the values in these two columns. Looking at the plot, do you think that the cash data follows a normal distribution?
   - In cell F2, enter =PERCENTRANK.EXC($B$2:$B$58, B2) and drag the formula to the remaining cells.
   - In cell G2, enter =NORMDIST(B2, $D4, $D5, TRUE) and drag the formula to the remaining cells.
   - Highlight the two columns. Click the Insert tab and then the Recommended charts icons. Select the scatterplot one.
   - In the Chart Tools tab in the top Excel menu, select the "Design" subtab. In the icon for "Quick Layout," select "Layout 3". In the x-axis type in "Actual PR" and in the y-axis type in "If Normal." Erase the Legends "If Normal" and "Linear(If Normal)."

   No, cash in pocket does not follow a normal distribution.

The results of these analyses are in the fourth worksheet of the Excel file for this chapter.

Random Numbers, Probability Distributions, Cross Tabulations

In the previous chapter, we mentioned that a good sample usually involves some type of random selection. In fact, a good sample is a ***random sample***, i.e., one in which each element in the population has the same chance to be selected into the sample. For example, in a class of 80 students we would like to obtain a random sample of 5. How do we do it? We can number each student from 1 to 80, enter their numbers in a hat, and pick five numbers out of the hat. We can replace this "physical" random selection by algorithms that generate ***pseudo-random numbers***, i.e., values that follow approximately a random or probabilistic pattern.

Also in the previous chapter, we introduce the normal distribution as a model for the distribution of a population. That distribution is one among several ***probability distributions*** that link the values of a distribution and the probability of their occurrence. Now, we will introduce two simple discrete distributions: the ***Bernoulli*** and the ***Binomial***. In contrast to the normal distribution, that can take continuous values, these two distributions take only discrete values, integers for example. The binomial is used for finding the probability of obtaining a given number of "successes" (whatever our definition is) out of a given number of attempts or trials.

Finally, we will introduce the use of ***cross-tabulations***, the frequency distribution of the joint values of two variables, to illustrate how to describe the occurrence of two events, and how the probability of occurrence of one event may be conditioned on the occurrence of the other event.

9.1 OBJECTIVES

1. To use random numbers procedures in Excel to obtain random samples from data sets.
2. To solve basic probability computations using Bernoulli or binomial distributions.
3. To obtain cross-tabulations using pivot tables.

9.2 RANDOM NUMBERS FUNCTIONS

Random numbers are values whose occurrence follows a probabilistic pattern. We can think of the tossing of a fair die as generating random integer numbers between 1 and 6, each of them with the same likelihood of occurrence. We can produce random numbers using mechanical methods of randomization, like in lotteries or bingos. On the other hand, we can use pseudo-random number generators that are algorithms to generate a sequence of random values. Excel has, among others, two functions for generating random numbers: the RAND and the RANDBETWEEN functions.

The RAND function generates a random decimal value between 0 and 1. Simply, type =RAND() in a cell to generate a random value. If you want to have more values, simply drag the function to additional rows.

The RANDBETWEEN function generates random integer values between a bottom and top value, each with the same probability of occurrence. For example, if you want to generate a random integer between 1 and 80, type in a cell =RANDBETWEEN(1, 80), and one value in that range will be generated in that cell. If you want more than one value, drag the function to as many cells as you want.

You will notice than when you drag a cell with a RAND() or RANDBETWEEN function, the values that the functions returned before change. The RAND and RANDBETWEEN are ***continuously active functions***, i.e., each time that you made any modification in the worksheet, these functions generate new random values. We actually can "refresh" the values produced by these two functions by pressing the F9 key.

If we want to keep a particular set of values produced by RAND() or RANDBETWEEN(), we need to copy and paste "*as values*" the results to another location:

- First, highlight the values and copy them to the clipboard (i.e., using Ctrl+C).
- While the range or original value is still highlighted, click the mouse right button and paste the content of the clipboard with the paste option "values."

In the first worksheet of the Excel file for this chapter, we use both functions to generate ten random values. In column A we have ten random values using RAND() and in column D we have ten random values between 1 and 80 using RANDBETWEEN (see Figure 9.1). Each time that we press the F9 key, the values in the two columns will change.

- In cell A5, enter RAND()
- Drag the function from cell A5 to cell A14
- In cell D5, enter RANDBETWEEN(1,80)
- Drag the function from cell D5 to cell D14

	A	B	C	D	E	F
1	Ten random values			Ten random values		
2	Using =RAND()			between 1 and 80		
3				Using =RANDBETWEEN		
4						
5	0.470405			23		
6	0.687833			57		
7	0.95326			2		
8	0.56985			70		
9	0.167293			18		
10	0.683169			73		
11	0.559035			58		
12	0.897496			56		
13	0.981587			28		
14	0.421973			13		
15						

Figure 9.1.

	A	B	C	D	E	F
1	Cash			Sample 1	Sample 2	sample 3
2	0			11	540	40
3	0			0	45	0
4	0			20	9	21
5	0			37	540	15
6	47			280	0	0
7	0					
8	280		Mean	69.6	226.8	15.2
9	15		SD	118.3947	286.4065	16.66433
10	7					

Figure 9.2.

Selecting a Random Sample of Values

The RANDBETWEEN function together with the INDEX and ROWS function allow us to select a random sample of values from a range of values in a worksheet.

For example, in the second worksheet in the Excel file for this chapter, we have in column A the amount of cash in pocket for 57 students (see Figure 9.2). We name this range (A2:A58) as "cash." We want to select a random sample of five observations from the "cash" range:

- To select the first random sample, enter in cell D2, the following formula:

  =INDEX(cash, RANDBETWEEN(1, ROWS(cash)))

The ROWS(cash) function returns the number of values in the range "cash," in our case 57. Thus, the RANDBETWEEN function returns a random value between 1 and 57. Finally, the INDEX function returns the value in the range "cash" located in the position indicated by the value returned by the RANDBETWEEN statement.

- To obtain the values on the other four cells, we drag the function in D2.
- In D8 we obtain the average of the sample, and in D9, the standard deviation.

We repeat the same procedure to obtain the other two samples in columns E and F. Of course, each time that we press the F9 key, the samples will change and also their means and standard deviations.

RANDOM VARIABLES AND PROBABILITY DISTRIBUTIONS

9.3

A ***random variable*** is a variable whose values, or interval of values, follow a probability distribution. A normal random variable is a ***continuous*** variable, because its values can take any real number, i.e., any value with decimals. In a continuous distribution are *intervals* of values that have probabilities. For example, in a standard normal distribution, the probability of obtaining a Z score of 1.96 or more is approximately 0.025. We will see later other continuous distributions such as the t- and F-distribution.

Discrete probability distributions are functions that link *individual* values with their probability of occurrence. They are called ***discrete*** because their values are countable values, i.e., integers. For example, tossing a fair die defines a discrete probability distribution because each of the six possible values has a probability of occurrence of 1/6. As with the normal distribution, discrete probability distributions play an important role in data analysis. The most common discrete random variables are the ones following a Bernoulli or a binomial distribution.

Bernoulli Distribution (Success vs. Failure)

Bernoulli is the simplest discrete random variable. This random variable has only two possible outcomes: A conventionally called "success" outcome that is represented by a value of 1. The other outcome is called a "failure" and it is represented by a value of 0. The probability of obtaining a value of 1 is represented with the letter p, while the probability of obtaining a value of 0 is represented with the letter q or by the expression $1 - p$. The result of tossing a coin is the classical example of data described by a Bernoulli distribution. If the coin is a fair one, the probability of a "success" (head or tail) is 0.5. We can use an Excel random number generator to "simulate" tossing coins. If you repeat the examples present below, you may find slightly different results because your randomly generated values will be different.

Suppose that we want to simulate tossing one fair coin (i.e., the probability of success $p = .5$). The value of the Bernoulli random variable X will be 1 if we obtain a tail (our "success") and 0 if we obtain a head (our "failure"). We simulate 1,000 tosses of this fair coin, and find the mean and standard deviation of the 1,000 simulations.

In worksheet 3 of the Excel file for this chapter, we have the simulated results (see Figure 9.3).

- We click on the Data tab in the top Excel menu, and then click on the Data Analysis icon. Select the "Random Number Generation" option in the Analysis Tool menu and click OK.
- In the Random Number Generation menu, enter 1 in the "Number of Variables" box, 1,000 in the "Number of Random Numbers" box, and select "Bernoulli" in the "Distribution" box. In the "p Value=" box enter .5. In the "Output Range" box enter a range starting at cell A3, and press OK. We can label this column as "Simulating 1000 Bernoulli."
- Compute the mean and standard deviation of the generated values. Put the mean in D3 and the standard deviation in D4. To facilitate computations, we name the range from A3 to A1002 as "simvalue." Notice that the mean is close to the expected probability of .5.

Binomial Distribution (Number of Successes in N Trials)

The binomial distribution describes the sum of N Bernoulli distributions. In other words, a binomial random variable is the number of "successes" (or scores of 1) that we obtain in N trials of Bernoulli random variables. The parameters of the binomial distribution are the number of trials N, and the common probability of "success" in each trial, p. Thus, the possible values of the binomial variable are integers between 0 (if there is no success at all) and N (if all the trials were successful).

	A	B	C	D	E	F	G	H	I	J
1	Simulating 1000					Simulating 1000				
2	Bernoulli					Binomial				
3	1		Mean	0.48		3		Mean	2.937	
4	1		SD	0.49985		3		SD	1.217585	
5	0					0				
6	1					3				
7	0					0				
8	1					4				

Figure 9.3.

Suppose that we ask a person to toss a fair coin $N = 6$ times and count the number of heads he/she obtains. We define as "success" obtaining a head. Because the coin is a fair one, the probability of obtaining a head is $p = .5$. Thus, the random variable "number of heads (successes)" can take values between 0 and 6. We will simulate 1,000 individuals performing the experiment. In column F of worksheet 3 in the Excel file for this chapter, we have the results.

- We click on the Data tab, and then on the Data Analysis icon. Select the "Random Number Generation" option in the Analysis Tool menu and click OK.
- In the Random Number Generation menu, enter 1 in the "Number of Variables" box, 1,000 in the "Number of Random Numbers" box, and select "Binomial" in the "Distribution" box. Enter .5 in the "p Value" box, and 6 in the "Number of trials" box. Select "Output Range" and enter a range starting at cell F3, and then press OK.
- Compute the mean and standard deviation for the generated values. Put the mean in I3 and the standard deviation in I4. To facilitate computations, we name the range from A3 to A1002 as "binvalue."

According to the binomial distribution, we expect to find an average of three "heads" when tossing six times a fair coin. In fact, the average of the 1,000 simulations is close to 3.

Probabilities in a Binomial Distribution

As with the case of the normal distribution, we can find the cumulative probabilities (percentile rank) in a binomial distribution. But in addition, we can find the probabilities for a single value in the distribution. We have three cases:

=BINOM.DIST(A1, 10, 0.25, FALSE)	It returns the ***actual probability for the value*** in cell A1 in a binomial with $N = 10$ and $p = .25$.
=BINOM.DIST(A1, 10, 0.25, TRUE)	It returns the ***cumulative probability for the value*** in cell A1 or a smaller value in a binomial with $N = 10$ and $p = .25$.
=BINOM.INV(10, 0.25, A1)	It returns ***the value*** defining the cumulative probability in cell A1 in a binomial with $N = 10$ trials and $p = .25$

With these functions, you can build tables of the binomial distribution or simply find probabilities for values that follow a binomial distribution.

For example, in the previous example of tossing six times a fair coin, what will be the probability of obtaining one success (head), two successes, etc.? In the fourth worksheet in the Excel file for this chapter, we get the table for a binomial distribution with $N = 6$, $p = .5$ (see Figure 9.4):

	A	B	C	D
1	Binomial N=6, p=.5			
2				
3	X	P(X)	P(X<x)	
4	0	0.015625	0.015625	
5	1	0.09375	0.109375	
6	2	0.234375	0.34375	
7	3	0.3125	0.65625	
8	4	0.234375	0.890625	
9	5	0.09375	0.984375	
10	6	0.015625	1	
11				
12				
13	Get four "six" out of 10 trials			
14	0.05457			
15				
16	Five corrects or less out of 20 items			
17	0.020695			
18				

Figure 9.4.

- We enter in column A (cells A4–A10) the possible values of X or "successes," in this case from 0 to 6.
- In cell B4, we enter BINOM.DIST(A4, 6, 0.5, FALSE). This is the probability, $P(X)$, of obtaining the corresponding value in A4.
- Drag the function to the remaining rows, up to cell B10.
- In cell C4, we enter BINOM.DIST(A4, 6, 0.5, TRUE). The returned value is the cumulative probabilities, $P(X < x)$, of obtaining the corresponding value in A4, or a smaller value.
- Drag the function to the remaining rows, up to cell C10.

We do not need to create a whole table to obtain a particular binomial probability or value. For example, we can use the binomial functions to answer questions like:

What is the probability of getting four "six" if we toss a fair die ten times? In this case a binomial with $N = 10$, $p = 1/6 = .167$ and $X = 4$.

- In cell A14, we enter =BINOMDIST(4, 10, 0.167, FALSE). The probability is 0.05457.

What is the probability of guessing correctly five or less questions in a test with 20 T/F items? In this case, a binomial with $N = 20$, $p = .5$, and $X = 5$.

- In cell A17, we enter =BINOMDIST(5,20,0.5,TRUE). Remember, "TRUE" refers to "true cumulative." The probability is 0.020695.

CROSS TABULATION AND MARGINAL, JOINT, AND CONDITIONED PROBABILITIES

9.4

The probability of the occurrence of a given event can be estimated from empirical data by observing the frequency of occurrence of the event and computing percentages. For example, a college professor would like to know the probability of having students majoring in communication disorders in her class. She surveys her class and finds the percentage of her students with that major. In worksheet 5 in the Excel file for this chapter, we have information about major, school year, age, and gender for 74 students. To find the percentage of students by major, we can use a pivot table to obtain a simple frequency or percentage distribution table (see Figure 9.5).

- Highlight column A (the major). Click on the Insert tab in the top Excel menu, and then click on the Pivot Table option.
- Choose "Existing Worksheet" as destination for the table, enter F1 for the output location of top-left corner of the table, and then click OK.
- Drag the major variable to the rows box. Drag the major variable again, but to the Values box. By default, the function will be Count. By default too, Excel introduces a "(blank)" entry. Click on the arrowhead next to "Raw Labels" and unselect the (blank) option.
- Right click the mouse on any of the counts values in the table, and select the "Value Field Settings" option. Click on the "Show value as" tab and select the option "% column total," and then click OK.

	A	B	C	D	E	F	G	H
1	Major	Year	Age	Gender		Row Labels	Count of Major	
2	Ag Econ	Junior	20	Male		Ag Econ	8.11%	
3	Ag Econ	Junior	20	Male		Communication disorders	12.16%	
4	Ag Econ	Junior	24	Male		HRMT	6.76%	
5	Ag Econ	Sophomore	19	Female		Psychology	58.11%	
6	Ag Econ	Sophomore	20	Male		Social Work	14.86%	
7	Ag Econ	Sophomore	20	Male		**Grand Total**	**100.00%**	
8	Communication disorders	Junior	20	Female				

Figure 9.5.

	A	B	C	D	E	F	G	H	I
11	Communication disorders	Junior	20	Female					
12	Communication disorders	Senior	21	Female		**Count of Major**	**Column Labe**		
13	Communication disorders	Senior	21	Female		**Row Labels**	**Female**	**Male**	**Grand Total**
14	Communication disorders	Sophomore	19	Female		Ag Econ	1.35%	6.76%	8.11%
15	Communication disorders	Sophomore	19	Female		Communication disorders	10.81%	1.35%	12.16%
16	Communication disorders	Sophomore	19	Male		HRMT	5.41%	1.35%	6.76%
17	HRMT	Junior	20	Female		Psychology	36.49%	21.62%	58.11%
18	HRMT	Junior	22	Male		Social Work	9.46%	5.41%	14.86%
19	HRMT	Senior	22	Female		**Grand Total**	**63.51%**	**36.49%**	**100.00%**
20	HRMT	Senior	22	Female					
21	HRMT	Sophomore	19	Female					

Figure 9.6.

Thus, 12.16% of the class are communication disorders majors. If we select at random one student from this class, the (empirical) probability that the student is a communication disorders major is therefore 0.1216. Because this probability refers to a single feature (the type of major), we call it a ***marginal probability***.

However, in many circumstances we would like to know the chances of two or more events happening at the same time. For example, the college professor would like to know the probability of having students that are communication disorders majors **and** female. Now we need a ***cross-tabulation table***, i.e., a table that counts the joint occurrence of the two events: type of major and gender of the student. Again, we use a pivot table (see Figure 9.6):

- Highlight all the columns (A–D). Click on the Insert tab in the top Excel menu, and then click on the Pivot Table option.
- Choose "Existing Worksheet" as destination for the table, enter F12 for the output location of the top-left corner of the table, and then click OK.
- Drag the major variable to the rows box. Drag the gender variable to the columns box. In the pivot table, click on the arrow icon for the row and unselect the (blank) option.
- Drag the major variable again, but to the Values box. By default, the function will be Count.
- Right click the mouse on any of the counts values in the table, and select the "Value Field Settings" option. Click on the "Show value as" tab and select the option "% Grand Total," and then click OK.

	A	B	C	D	E	F	G	H	I	J
22	Psychology	Freshman	20	Female						
23	Psychology	Freshman	19	Male		**Count of Major**	**Column Labels**			
24	Psychology	Junior	20	Female		**Row Labels**	**Female**	**Male**	**Grand Total**	
25	Psychology	Junior	20	Female		Ag Econ	2.13%	18.52%	8.11%	
26	Psychology	Junior	20	Female		Communication disorders	17.02%	3.70%	12.16%	
27	Psychology	Junior	20	Female		HRMT	8.51%	3.70%	6.76%	
28	Psychology	Junior	20	Female		Psychology	57.45%	59.26%	58.11%	
29	Psychology	Junior	21	Female		Social Work	14.89%	14.81%	14.86%	
30	Psychology	Junior	21	Female		**Grand Total**	**100.00%**	**100.00%**	**100.00%**	
31	Psychology	Junior	21	Female						
32	Psychology	Junior	22	Female						

Figure 9.7.

The table we have created is referred as a ***joint probability table*** (although we use here percentages instead of proportions). We read that 10.81% of all students in the class are female **and** communication disorders majors. In other words, if we select at random one student from this class, the (empirical) probability that the student is a communication disorders major **and** a female is 0.1081. Because this probability refers to two features (major **and** gender), it is referred to as the ***joint probability*** of being "communication disorders" major and female. Notice that the last column of the table (the Grand Total) contains the "marginal" probability for each of the majors.

The probability of a given event sometimes is affected by the actual occurrence of another event. For example, in the table for the marginal probability of majors we have that the probability of being a Psychology major is 0.5811, or 58.11% (see Figure 9.7). We wonder if knowing the gender of the student will affect that probability. Maybe if we know that the student is female, the chances of being a Psychology major will increase. We need a cross-tabulation table, where the percentages are computed by using the total frequency of each column (i.e., percentages by each gender). The easiest way to obtain this table is to copy our previous pivot table and modified the direction of the percentages.

- Highlight and select the whole pivot table and copy it to the buffer (Ctrl + C).
- Select cell F23 and paste the pivot table.
- Right click the mouse on any of the counts values in the table, and select the "Value Field Settings" option. Click on the "Show value as" tab and select the option "% of Column Total," and then click OK.

The table shows percentages by columns. Thus, out of all female students, 57.45% of them are Psychology majors. On the other hand, out of all male students, 59.26% of them are Psychology majors. Another way to state the same information emphasizes the conditioning of the probability by the occurrence of an event: The probability of being a Psychology major **given that** the student is female is 0.5745. Similarly, the probability of being a Psychology major **given that** the student is male is 0.5926. Because these probabilities are conditioned on knowing the gender of the student, they are called ***conditioned probabilities***.

Of course, we can change the direction of the percentages by selecting as option "% of Row Total." In this case, the percentages will give us a different conditioning, i.e., the probability that a student is male or female, **given** that we know his/her major. For example, the probability of a student being female given that her major is Psychology is 0.6279.

9.5 WORKED EXAMPLE

We use the sixth worksheet in the Excel file for this chapter to obtain the following analyses (see Figure 9.8).

1. According to probability theory, the ***average*** of two Bernoulli random variables with $p = .5$ should have a mean of .5 and a standard deviation of 0.3536. Simulate 100 values for each of the two Bernoulli, find their average and obtain the mean

	A	B	C	D	E	F	G	H	I	J	K	L
1	Bern 1	Bern 2	Mean			Year	Gender		**Count of Year**	**Column Labels**		
2	0	0	0			Junior	Male		**Row Labels**	**Female**	**Male**	**Grand Total**
3	0	1	0.5	Mean	0.515	Junior	Male		Freshman	1.35%	1.35%	2.70%
4	1	1	1	SD	0.358553491	Junior	Male		Junior	22.97%	12.16%	35.14%
5	0	1	0.5			Sophomore	Female		Senior	8.11%	2.70%	10.81%
6	1	0	0.5			Sophomore	Male		Sophomore	31.08%	20.27%	51.35%
7	0	0	0			Sophomore	Male		**Grand Total**	**63.51%**	**36.49%**	**100.00%**
8	0	1	0.5			Junior	Female					
9	1	1	1		40 or less imrpoved	Junior	Female		**Count of Year**	**Column Labels**		
10	1	0	0.5		0.028443967	Junior	Female		**Row Labels**	**Female**	**Male**	**Grand Total**
11	1	1	1			Junior	Female		Freshman	2.13%	3.70%	2.70%
12	0	0	0		More than 65 improved	Senior	Female		Junior	36.17%	33.33%	35.14%
13	0	0	0		0.000894965	Senior	Female		Senior	12.77%	7.41%	10.81%
14	1	0	0.5			Sophomore	Female		Sophomore	48.94%	55.56%	51.35%
15	0	1	0.5		Exactly 50 improved	Sophomore	Female		**Grand Total**	**100.00%**	**100.00%**	**100.00%**
16	0	0	0		0.079589237	Sophomore	Male					
17	1	0	0.5			Junior	Female					
18	0	0	0			Junior	Male					
19	1	1	1			Senior	Female					

Figure 9.8.

and standard deviation of the 100 means. Are the mean and standard deviation close to what is expected?

- Select the Data tab in the top Excel menu, then click on the Data Analysis ToolPak item, and select the Random Number Generation option.
- In the Random Number Generation menu, enter 2 as "Number of Variables" and 100 as "Number of Random Numbers." Select as distribution Bernoulli with p-value = .50. Select "Output Range" and enter A2 as the first column for the output.
- In cell C2, enter =AVERAGE(A2:B2) and drag the function to the remaining cells.
- Finally, enter in E3 the mean and in E4 the standard deviation of the values in column C.

Yes, the simulated mean is close to 0.50 and the simulate standard deviation is close to 0.3536.

2. Suppose that in a class of $N = 100$ students, the instructor decides to label a student as "improved" if he/she gets a higher score in the second exam than in the first exam. Suppose that the chance that a student increases in the second exam is $P(\text{Increase}) = p = .5$. In consequence, the number of "improved" students will follow a binomial distribution with $N = 100$ and $p = .5$
   a) According to this binomial distribution, what is the probability of observing 40 or less "improved" students in the class?
      - In E9, enter BINOM.DIST(40, 100, 0.5, TRUE). The outcome is P(X<= 40) = .028444.
   b) According to this binomial distribution, what is the probability of observing more than 65 "improved" students in the class?
      - In E13, enter 1-BINOM.DIST(65, 100, 0.5, TRUE). The outcome is P(X> 65) = .000895.
   c) According to this binomial distribution, what is the probability of observing exactly 50 "improved" students in the class?
      - In E16, enter BINOM.DIST(50, 100, 0.5, FALSE). The outcome is P(X = 50) = .079589.

3. In columns F and G, we have the class year and the gender of 74 students. Obtain a cross-tabulation for class year on the rows and gender in the columns. If we select at random a student from this table
   a) What is the probability that the student is a junior?
   b) What is the probability of being a junior and a female?
   c) If we know that the student selected is male, what is the conditional probability that the student is a junior?
      - Highlight columns F and G. Click on the Insert tab in the top Excel menu, and then click on the Pivot Table icon. Enter I1 for the output location of the table, and then click OK.
      - Drag Year to the Rows box. Drag Gender to the Columns box. Drag Year (or Gender) to the Values box. By default, the function will be Count.
      - Right click the mouse on any of the counts values in the table, and select the "Value Field Settings" option. Click on the "Show value as" tab and select the option "% Grand Total," and then click OK.

   d) The probability of being a junior is 0.3514 (or 35.14%)
   e) The probability of being a junior and a female is 0.2297 (or 22.97%)
      - Highlight and select the whole pivot table and copy it to the buffer (Ctrl + C).
      - Select cell I9 and paste the pivot table.
      - Right click the mouse on any of the counts values in the table, and select the "Value Field Settings" option. Click on the "Show value as" tab and select the option "% of Column Total," and then click OK.

   f) The probability that the student is a junior given that the student is male is 0.3333 (or 33.33%)

The final results are in worksheet 7 in the Excel file for this chapter.

10 Sampling Distribution of the Mean and Confidence Intervals

When analyzing data it is unusual that we know for sure the shape of the distribution of values in the population of interest. We usually make some assumptions, such as that the data is normally distributed. In this chapter, we introduce inferential principles that not only support the use of normal distribution, but also provide us with a practical procedure to estimate the mean of a population.

Suppose that our population of interest comprises the grade point average (GPA) of all four-year college students in the nation. We would like to know their population GPA average μ. They are several millions of these students and, of course, we do not know the distribution of their GPA. However, with some effort, we are able to obtain a *single* good national random sample of $N = 1{,}600$ students' GPA. How can we use these sample data to estimate the population average GPA?

We need to reformulate our task by introducing a new distribution: The distribution of sample means. Let us start with a mental experiment: Imagine that instead of only one *single* sample from our target population, we obtain a very large number of samples of the same size $N = 1{,}600$ (in fact, the theory argues for *all* the possible samples of this size). In each one of these samples, we compute the sample mean $\overline{X}$ and plot the distribution of all these sample means. This distribution of means is the ***sampling distribution of the mean***. What is outstanding is that the shape of this distribution of means approaches the shape of a normal distribution when the sample size increases. This property is referred to as the ***central limit theorem (CLT)***. Therefore, in our example of samples of size $N = 1{,}600$ the distribution of sample means definitely will take a normal shape.

Imagine now that we obtain the average of all the sample means $\overline{X}$ in the sampling distribution. The mean of all the sample means will be equal to the mean of the population from where the samples came. In other words, if we keep sampling and computing $\overline{X}$, in the long run (on average) we will obtain the value of the population mean μ.

Of course, the sample means, $\overline{X}$, are not going to take the same value in all samples. Thus, the standard deviation of the sampling distribution of the means tells me the spread of possible values for the sample means. This standard deviation receives the new name of ***standard error of the mean***, and it is a function of the population standard deviation and the sample size.

Going back to our GPA question, we know now by the CLT that our single sample mean is a member of a normal distribution whose mean is the population mean we want to know. In addition, we know that the standard deviation (or standard error) of the distribution is a function of the population standard deviation and the sample size. Based on these, we create an interval of values around our sample mean that may contain the actual population mean value. These intervals are called ***confidence intervals***, where the confidence that the interval contains the actual population mean is expressed as a percentage.

10.1 OBJECTIVES

1. To describe the properties of the sampling distribution of the mean.
2. To compute confidence intervals using the standard normal distribution.

10.2 GENERATING A SAMPLING DISTRIBUTION OF THE MEAN

The ***sampling distribution of the mean*** is the distribution of the means from all possible samples from a target population. When the size of the sample, i.e., the number of values in the sample, increases: (1) the shape of the sampling distribution approaches the normal distribution, and (2) the standard deviation of the sampling distribution, called ***standard error***, becomes smaller.

We will illustrate the generation of the sample distribution by running a simulation in Excel. We will use as the original "unknown" population the discrete distribution of integers between 0 and 9. Each integer value in this distribution has the same probability of being selected, i.e., 1/10 or 0.1. The values can be generated using the =RANDBETWEEN(0, 9) function. The ***expected value***, or average, and the standard deviation for this population are $\mu = 4.5$ and $\sigma = 2.87$. The histogram for this population will be "flat," i.e., all the bars will have the same height.

According to the theory, the mean and the standard deviation (or standard error) of the sampling distribution of the sample means, $\overline{X}$, for samples of size N is:

$$\mu_{\overline{X}} = \mu \qquad \sigma_{\overline{X}} = \sigma/\sqrt{N}$$

Thus, for a sample size of $N = 5$, the mean of the sampling distribution will be $\mu_{\overline{X}} = 4.5$, i.e., equal to the population mean. The standard error will be $\sigma_{\overline{X}} = 2.87/\sqrt{5} = 1.285$, i.e., the population standard deviation divided by the square root of the sample size.

In the first worksheet in the Excel file for this chapter, we illustrate the generation of a sampling distribution of the mean (see Figure 10.1). We generate 100 samples. Each row contains the five values in a sample and the sample average.

- In column A, we enter an ID number for the sample. Enter 1 in A1 and 2 in A2. Select both cells and drag them to generate the remaining ID numbers.
- In columns B–F, we will simulate the five values for each sample (X1–X5). In B2 enter =RANDBETWEEN(0,9).
- Drag cell B2 up to cell F101. An array of columns. The function will repeat for all the entries in the rectangle array.
- In cell G2, enter =AVERAGE(B2:F2). This will be the average of the first sample.
- Drag cell G2 to cell G101. The means for all the samples.
- In cell K2, enter =AVERAGE(G2:G101). This is the mean of all the means.
- In cell K3, enter =STDEV.S(G2:G101). This is the standard error of all the means.
- Create the histogram for the means in column G: highlight column G; select the Insert tab in the Top Excel menu and in the Charts section select the histogram. Once the graph is plotted, edit it to have an interval width of 0.5

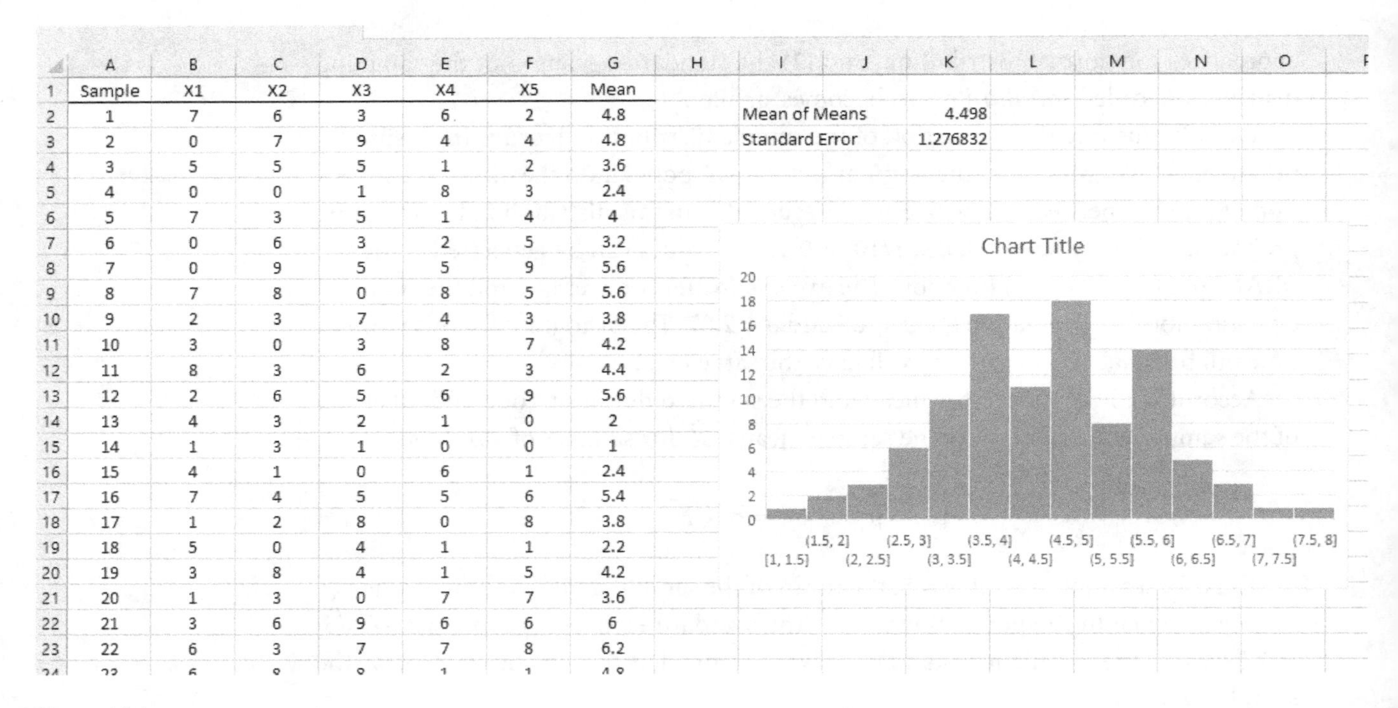

Sample	X1	X2	X3	X4	X5	Mean
1	7	6	3	6	2	4.8
2	0	7	9	4	4	4.8
3	5	5	5	1	2	3.6
4	0	0	1	8	3	2.4
5	7	3	5	1	4	4
6	0	6	3	2	5	3.2
7	0	9	5	5	9	5.6
8	7	8	0	8	5	5.6
9	2	3	7	4	3	3.8
10	3	0	3	8	7	4.2
11	8	3	6	2	3	4.4
12	2	6	5	6	9	5.6
13	4	3	2	1	0	2
14	1	3	1	0	0	1
15	4	1	0	6	1	2.4
16	7	4	5	5	6	5.4
17	1	2	8	0	8	3.8
18	5	0	4	1	1	2.2
19	3	8	4	1	5	4.2
20	1	3	0	7	7	3.6
21	3	6	9	6	6	6
22	6	3	7	7	8	6.2

Mean of Means	4.498
Standard Error	1.276832

Figure 10.1.

Notice that each time that we work on the spreadsheet, the values change. This is because the RANDBETWEEN is an active function that generates a new set of values each time that there is a modification in the spreadsheet. We can also "refresh" the output with a new set or sample values by pressing the function key F9. If you refresh the values, you will see that the "Mean of means" fluctuates around 4.5, as we expect based on the theory. The value of the standard error will fluctuate around 1.285. The shape of the distribution will change too, but it will tend to resemble a normal distribution. In order to have a more normal shaped distribution, we will need to have samples larger than $N = 5$.

10.3 CONFIDENCE INTERVAL FOR THE MEAN (USING STANDARD SCORES Z)

A ***confidence interval*** is an estimator for a population parameter that provides an interval of values to which we assign some "confidence" of containing the value of the population parameter. In the case of the population mean, if we know the population standard deviation, σ, we build a confidence interval by using our sample mean and the standard normal distribution (the *Z*-distribution). Suppose now that we obtain a large number of samples (actually all samples of a given size) and compute the confidence interval for the mean in each sample, then the ***confidence level*** is the percentage of intervals that actually include the population mean. The confidence level guides the computation for the upper and lower limits of the interval. Because the sampling distribution of the mean follows a normal distribution:

90.00% of the values are between *Z* scores of approximately −1.64 and +1.64

95.00% of the values are between *Z* scores of approximately −1.96 and +1.96

99.00% of the values are between *Z* scores of approximately −2.58 and +2.58

Then, the 90%, 95%, and 99% confidence interval for the population mean, when we know the population standard deviations are given by:

$$\overline{X} \pm 1.64\left(\sigma/\sqrt{n}\right) \qquad \overline{X} \pm 1.96\left(\sigma/\sqrt{n}\right) \qquad \overline{X} \pm 2.58\left(\sigma/\sqrt{n}\right)$$

In general, the 100*(1 – α)% confidence interval formula is

$$\overline{X} \pm Z_{\alpha/2}\left(\sigma/\sqrt{n}\right)$$

where the term $Z_{\alpha/2}$ refers to the *Z* scores that define the desired percent confidence. Notice that ***alpha***, α, is the proportion that is not covered in the interval. Thus, for a 95% confidence interval (or 0.95) there are 5% not covered, and this percentage is a proportion of α = .05.

Excel has the function CONFIDENCE.NORM that produces the value $Z_{\alpha/2}\left(\sigma/\sqrt{n}\right)$, once we provide the desired alpha, the standard deviation, and the sample size. This value is equal to half the width of the confidence interval. In worksheet 2 in the Excel file for this chapter, we have an example of how to use this function to obtain the 95% confidence interval for the mean (see Figure 10.2):

- We know that a population is normally distributed with a standard deviation σ = 4. In column B, we enter a random sample of size of five elements: 99, 97, 100, 107, 101.
- In cell B7, we obtain the mean =AVERAGE(B2:B6).

	A	B
1		Data
2		99
3		97
4		100
5		107
6		101
7	Mean	100.8
8		
9	Half-width	3.506090162
10		
11	95% Lower	95% Higher
12	97.29390984	104.3060902
13		

Figure 10.2.

- We put the confidence interval half-width in B9 using =CONFIDENCE.NOM(.05, 4, 5). The value .05 refers to 5% not covered, the value 4 is the population standard deviation, and 5 is the sample size. The reported half-width of 3.5060902.
- In cell A12, we enter the sample mean minus the half-width, (B7 – B9), i.e., the lower limit for the 95% confidence interval.
- In cell A13, we enter B7+B9, i.e., the upper limit for the 95% confidence interval.

Thus, we estimate that the population mean is between 97.29 and 104.31, with 95% confidence.

10.4 WORKED EXAMPLE

The width of a confidence interval relates to the "precision" of the estimate. Of course, the precision is a function of (a) the desired level of confidence (i.e., 95% or alpha of .05, vs. 99% or alpha .01), (b) the standard deviation of the population (i.e., 4 vs. 12), and of course (c) the sample size (i.e., 30 vs. 100).

1. Using the CONFIDENCE.NOM function obtain a table for the half-width of the confidence interval using the confidence levels, standard deviation, and sample sizes mentioned above (see Figure 10.3).
    - In worksheet 3 in the Excel file for this chapter, we enter alpha in column A, standard deviations in column B, and sample sizes in column C.
    - From cell A3–A6 enter the value .01. From cell A7–A10 enter .05.
    - From cell B3–B10 enter 4,4,12,12, 4,4,12,12.
    - From cell C3–C10 enter 30,100,30,100,30,100,30,100.
    - In cell D3, we enter the function CONFIDENCE.NORM(A3, B3, C3) to obtain the half-width for alpha .01 (or 99%), standard deviation 4, and sample size 30. Drag the function to the remaining cells.

2. Which combination, or combinations, of confidence level, standard deviation, and sample size produced the smallest half-width, or best precision?

   The best combination is a 95% confidence with standard deviation 4 and sample size 100. The half-width is 0.784 in cell D8.

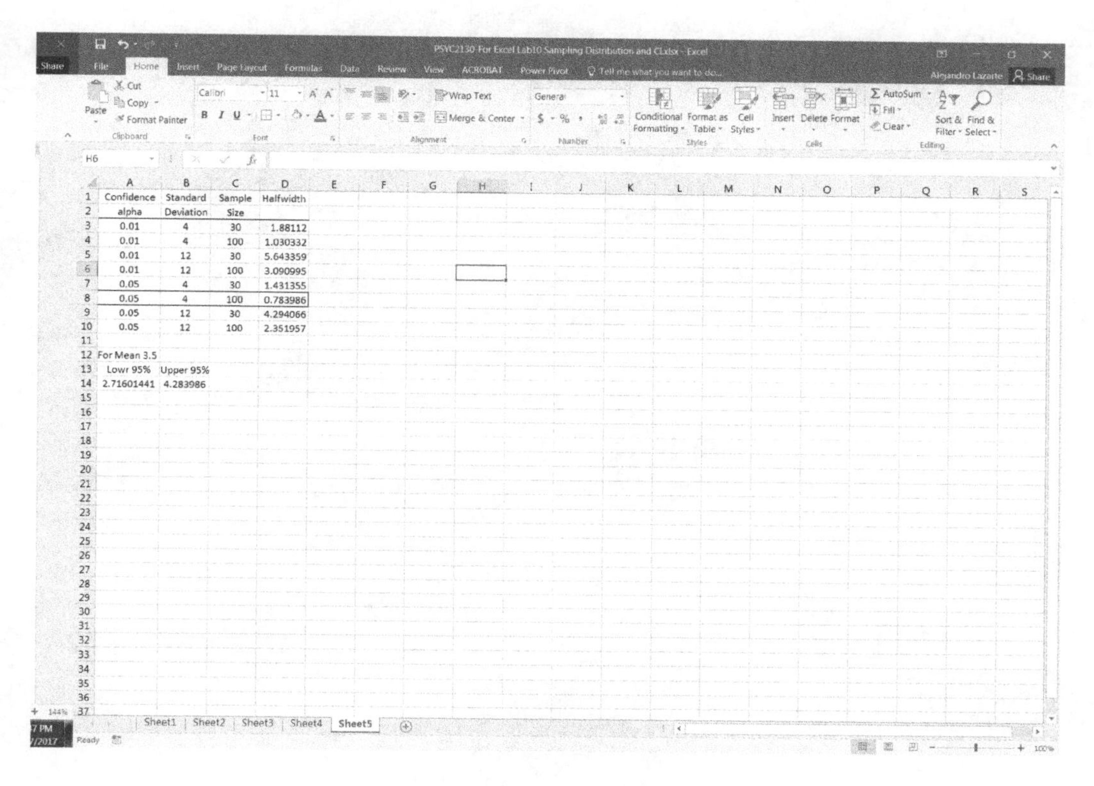

	A	B	C	D
1	Confidence	Standard	Sample	Halfwidth
2	alpha	Deviation	Size	
3	0.01	4	30	1.88112
4	0.01	4	100	1.030332
5	0.01	12	30	5.643359
6	0.01	12	100	3.090995
7	0.05	4	30	1.431355
8	0.05	4	100	0.783986
9	0.05	12	30	4.294066
10	0.05	12	100	2.351957
11				
12	For Mean 3.5			
13	Lowr 95%	Upper 95%		
14	2.71601441	4.283986		

Figure 10.3.

3. Suppose that the sample mean is 3.5, report the confidence interval using the best precision
   - To obtain the lower limit, enter in cell A14, the expression =3.5-D8.
   - To obtain the upper limit, enter in cell B14, the expression =3.5+D8.

   [2.71601441, 4.283986]

The final output is in worksheet 4 in the Excel file for this chapter.

11 One-Sample Z-Test: p-Value, Effect Size, and Power

In Chapter 10, we introduced the sampling distribution of the mean as our model for estimating the population mean using a confidence interval (CI). In this chapter, we introduce the other side of inferential statistics: ***hypothesis testing***. In statistical hypothesis testing, we advance a hypothesis about the value of a population parameter and we assess how likely our sample statistic is if that hypothesis is true. Of course, we can test different population parameters, but we will focus now on the population mean. Thus, we asses if a sample mean can be considered as a likely member of a sampling distribution whose mean is stated by the null hypothesis. For example:

> Female average height in the United States is $\mu = 65$ inches with a standard deviation of $\sigma = 3.5$ inches. We are given the heights of a sample of $N = 36$ females. The sample mean height is $\overline{X} = 67.5$. How likely is it to obtain this value of sample mean, or a more extreme value, if our sample is a random sample from the general female population of the United States? Alternatively, can we consider the sample mean a likely member of the sampling distribution defined by a population mean of $\mu = 65$?

A ***statistical hypothesis*** is simply a statement about the value of a population parameter. We distinguish two types of statistical hypothesis:

The ***null hypothesis*** is the statement that states the value that we assume that the population parameter takes. We use this value in further computations. On the other hand, the ***alternative hypothesis*** refers to the logical counterpart of the null hypothesis. In our example above, we start assuming that the sample is a random sample from the general US female population. Therefore, the null hypothesis will state that the population average height is 65 inches. The alternative hypothesis will be the logical opposite, i.e., one alternative is that the population mean is not 65 inches. These two statements are represented as:

$$H_0: \mu = 65 \qquad H_1: \mu \neq 65$$

Thus, with this null hypothesis, sample size, and population standard deviation, we use a sampling distribution of the means with normal shape, mean $\mu = 65$ and standard error of $\sigma/\sqrt{N} = 3.5/\sqrt{36} = 0.583$. In this sampling distribution, where is a sample mean $\bar{X} = 67.5$? We locate the sample mean of 67.5 in the sampling distribution by finding its standard score in the distribution:

$$Z_{obt} = \frac{\bar{X} - \mu}{\sigma/\sqrt{N}}$$

Z_{obt} is a ***test-statistic***. A test-statistic is a value that relates statistics in our sample, i.e., our sample mean, with the value of the parameter postulated in the hypothesis, the value of μ. If the sample mean is too far away from the value in the hypothesis, Z_{obt} will take a large value in absolute terms. On the other hand, if the sample mean is close to the hypothesis, the value of Z_{obt} will be close to zero. In our example, the value of Z_{obt} is:

$$Z_{obt} = \frac{67.5 - 65}{3.5/\sqrt{36}} = 4.28$$

Hence, a sample mean of 67.5 is located 4.28 standard errors above the hypothesis of $\mu = 65$ in a normal distribution. This is an extreme location because in a normal distribution nearly all the observations are between plus and minus three standard deviations. Thus, this sample mean of 67.5 is very unlikely to be a member of the sampling distribution. We express how likely or unlikely is the sample mean by obtaining its ***p-value***:

> ***p-value is the probability of obtaining the value of the test-statistic, or a more extreme value, if the null hypothesis of the test is true.***

The "more extreme value" part in the previous statement refers to the directionality of the alternative hypothesis. In our case, our alternative was that the population mean was "different" from 65. This type of alternative hypothesis is called a ***two-tail alternative hypothesis*** or nondirectional hypothesis because it is open to results that differ from the null hypothesis in any direction, i.e., larger or smaller than the null hypothesis. Consequently, when computing the *p*-value we find the probability of obtaining a value of 4.28 or more (the "positive tail") or a value of −4.28 or less (the "negative tail"):

$$p-value = P(Z \leq -4.28) + P(Z \geq 4.28) = 0.000019$$

Thus, we have that: "The probability of obtaining a Z_{obt} of 4.28, or more extreme value in any direction, is 0.00019 if we assume that the sample comes from a population with mean 65."

When does the *p*-value indicate that the probability of the test-statistic under the null hypothesis is too unlikely? A standard criterion in statistical analysis is to use a ***critical value or alpha level value*** for deciding when to reject or retain a null hypothesis. The critical value is actually the probability of making the mistake of rejecting the null hypothesis when this hypothesis is actually true (called a Type I error). It is common practice to decide in advance to use as critical values .05 or .01. Thus, suppose that we decide that if the *p*-value is smaller than .01 we will reject the null hypothesis. In our example, we have:

$$p-value = 0.000019 \ < .01$$

Thus, we reject the null hypothesis that the sample comes from a population with mean 65 inches. The sample does not seem to be a random sample from the heights of general population of US females. The sample may be from a special US subpopulation, i.e., basketball players.

Rejecting a null hypothesis, i.e., finding that a sample statistic differs from a hypothesis, is referred to as finding a ***statistical significant difference***. However, the "significant" does not imply that the difference found is large. ***Effect size*** (*d*) is a measure of the magnitude of the actual difference found between observed statistic and null hypothesis values. In the case of *Z*-test for one sample, the effect size expresses the difference between the hypothesis and the observed sample mean in terms of the original population standard deviation. In our example:

$$d = \frac{67.5 - 65}{3.5} = .714$$

This effect size value, sometimes referred as ***Cohen's d***, is judged conventionally as follows:

d values around 0.2 are considered small effect size.
d values around 0.5 are considered medium effect size.
d values around 0.8 (or more) are considered large effect size.

In our example, the effect size is closer to the classification as large effect size.

The alpha level or critical value for deciding to reject or retain the null hypothesis attempts to control the probability of making a Type I error (i.e., rejecting the null hypothesis when it is actually true). However, another possible error when testing a null hypothesis is to fail to reject the null hypothesis when it is actually false. This is called a Type II error. In testing, we attempt to control this type of error by increasing the "power" of the test. The ***power of a test*** refers to the probability that the test rejects the null hypothesis when actually the null hypothesis is false. The power of a test depends on several features of the test such as the alpha level, the standard deviation of the response, the difference between the null and a specific alternative hypothesis, and the sample size. Most of the time, we increase sample size to obtain a better power for a test.

11.1 OBJECTIVES

1. To compute a Z-test for one sample using Excel basic functions and the formula for the test.
2. To use the chapter template to compute a Z-test for one sample, its effect size, and 95% CI for the mean.
3. To use the simple power template for a Z-test for one sample.

11.2 Z-TEST FOR ONE SAMPLE

In the first worksheet of the Excel file for this chapter, we have the scores for two samples of size $N = 6$ from a normal population with known standard deviation $\sigma = 5$. They are in columns B and C.

Sample 1: 97, 96, 100, 108, 103, 99
Sample 2: 83, 78, 80, 75, 82, 79

We want to test the null hypothesis that these samples come from a population with mean 100 vs. an alternative hypothesis that they are coming from a population with a mean different from 100, i.e.,

$$H_0: \mu = 100 \text{ vs.} \qquad H_1: \mu \neq 100$$

First, we find the means for each sample (see Figure 11.1).

- In cell B8, enter =AVERAGE(B2:B7). In cell B9, enter =COUNT(B2:B7)
- In cell C8, enter =AVERAGE(C2:C7. In cell C9, enter =COUNT(C2:C7)

	A	B	C	D
1		Sample 1	Sample 2	
2		97	83	
3		96	78	
4		100	80	
5		108	75	
6		103	82	
7		99	79	
8	Mean	100.5	79.5	
9	N	6	6	
10				
11	Z obt	0.24495	-10.043	
12	p-value	0.8065	9.9E-24	
13				
14	Effect Size	0.1	4.1	
15				

Figure 11.1.

Finding the Test Statistic: Z_{obt}

We compute the test-statistic, Z_{obt}, for each sample using the formula

$$Z_{obt} = \frac{\bar{X} - \mu}{\sigma / \sqrt{N}}.$$

- Enter in cell B11, the equation =(B8-100)/(5/SQRT(B9)). This will give the test-statistic for the first sample.
- Drag the equation in cell B11 to C11. This will give the test-statistic for the second sample.

Finding the p-Value for Two Tails

To obtain the *p*-value for the Z_{obt}, we have to consider the **alternative hypothesis.** In this example, the alternative hypothesis is "*not-directional*" or "*two tails.*" In other words, we do not know ahead of time if the Z_{obt} will be positive or negative, and therefore we allow for those two possible outcomes. The probability of "more extreme values" for Z_{obt}, therefore, involves ***both*** tails of the standard normal distribution. Because the standard normal distribution is symmetric, we can find the probability for one of the tails and multiple this probability by two. Because the normal distribution function in Excel provides areas below a given value of Z, we will use the case when Z_{obt} is negative (i.e., the lower tail of the distribution). However, because we do not know if the Z_{obt} is going to be positive or negative ahead of time, we can take the absolute value of the Z_{obt} and attach to that absolute value the negative sign. In this way, we assure that Z_{obt} is negative, and the probability in the normal distribution will be in the lower tail. Thus, the *p*-value for the two-tail test will be:

$$p - value = 2 \times P\left(Z < -abs\left(Z_{obt}\right)\right)$$

- We use the NORM.S.DIST function to obtain the *p*-value. In our example, enter in cell B12, the equation =2*NORM.S.DIST(-ABS(B11), TRUE) for the *p*-value for the first sample.
- Drag the equation to cell C12 to obtain the *p*-value for the second sample.

For the first sample, the *p*-value is .8065. This value is larger than an alpha level of .05. Therefore, we cannot reject the null hypothesis that this sample comes from a population with mean 100.

For the second sample, the *p*-value is in scientific notation: 9.9×10^{-24}. In other words, we move the decimal point 24 positions on the left, i.e., a very small number. Thus, the *p*-value in the second sample is definitely smaller than .05 and therefore we reject the null hypothesis that this sample comes from a population with mean 100.

Effect Size

The Cohen's *d* effect size is a number that attempts to measure the magnitude of the departure from the null hypothesis. It expresses the difference between the sample mean and the hypothesis mean in terms of the actual standard deviation of the population, i.e., $d = \frac{|\bar{X} - \mu|}{\sigma}$. The bars around the numerator represent the absolute value of the difference. In our example:

- Enter in cell B14, the expression = ABS(B8-100)/5. This is the effect size of the first sample.
- Drag the equation to the cell C14. This is the effect size of the second sample.

For sample 1, the effect size is quite small, i.e., only 0.1 standard deviations of the population. On the other hand, the magnitude of the effect size in sample 2 is quite large. Sample 2 mean is over four standard deviations from the hypothesis of $\mu = 100$.

A Template for Z-Test for One Sample

The computations for the Z-test for one sample can be easily put into a template. In the second worksheet of the Excel file for this chapter, we have an Excel template to perform the one-sample Z-test and obtain the Z_{obt}, *p*-value, effect size, and the 95% CI for the population mean. In the gray cells we enter as input for the template: (1) the value of the null hypothesis, (2) the direction of the alternative hypothesis, <> for a two-tail test, < for a one-tail lower test, or > for one-tail upper test, (3) the standard deviation of the population, (4) the obtained sample mean, and (5) the sample size.

In the template on the left (Figure 11.2), we enter the data for first sample and request a two-tail test. The ***p*-value** for this test is .806496. For a critical value of $\alpha = .05$, we cannot reject the null hypothesis that the sample mean comes from a population with mean 100. In fact, the 95% CI shows that 100 is inside the 95% CI for the mean.

However, the template on the right (Figure 11.3) shows a zero ***p*-value** for the two-tail test for the sample mean of 79.5. Actually, the *p*-value is not zero, but an extremely small value. Therefore, we reject the null hypothesis that the sample mean comes from a population with mean 100. Actually, the 95% CI estimates the population mean for this sample between 75.5 and 85.5 approximately, i.e., 100 is not one of those possible means. In addition, the effect size is quite large.

	A	B	C
1	Z test for Hypothesis about a single mean		
2			
3	Null Hypothesis H0	100	
4	Alternative	<>	(<>, <, >)
5			
6	Standard Deviation	5	
7			
8	Sample Mean	100.5	
9	Sample Size	6	
10			
11	Standard Error	2.041241	
12	Zobt	0.244949	
13	p-value	0.806496	
14			
15	Effect size	0.1	
16			
17	95% C.I.	96.49917	104.5008
18			

Figure 11.2.

	A	B	C
1	Z test for Hypothesis about a single mean		
2			
3	Null Hypothesis H0	100	
4	Alternative	<>	(<>, <, >)
5			
6	Standard Deviation	5	
7			
8	Sample Mean	79.5	
9	Sample Size	6	
10			
11	Standard Error	2.041241	
12	Zobt	-10.0429	
13	p-value	0	
14			
15	Effect size	4.1	
16			
17	95% C.I.	75.49917	83.50083
18			

Figure 11.3.

11.3 POWER

As we mentioned before, the **power of the test** is defined as the

> *Probability of rejecting the null hypothesis when the null hypothesis is actually false.*

The power is a function of (1) the difference between the value of the null hypothesis and the value of a ***specific*** alternative hypothesis, (2) the standard deviation of the population, (3) the desired alpha level, and (4) the sample size. In most applied settings, we look for a sample size that provides a high enough power for the test.

In the third worksheet in the Excel file for this chapter, we have a simple template for performing power computations for the one-sample Z-test for a two-tail test. There are actually two templates. The first one (to find POWER) computes the power given sample mean, sample size, alpha, and hypotheses. The second one (to find SAMPLE SIZE) computes the required sample size given that we provide the sample mean, alpha, and hypotheses.

Finding Power

We want to test the null hypothesis $\mu = 100$, for a population with standard deviation is $\sigma = 5$. We want to use an alpha level of $\alpha = .05$, and we have a sample size of $N = 6$. Suppose that we want to detect a change in the population mean of 0.5, i.e., we postulate as specific alternative hypothesis that the population mean is 100.5.

Then, after entering all the previous information in the POWER template, the template calculates a power of 0.0431712 (see Figure 11.4). Thus, the power for detecting a difference between a mean of 100 and 100.5 when using, among other things, a sample size of only six observations is too low. It seems that we need to increase the sample size if we want to have a higher power.

Finding Sample Size

When planning an experiment, we want to know the sample size that we need in order to obtain a good power for the test. In the previous example, we got a very low power when using a sample of size 6. Now, keeping all the same alpha and hypotheses

	A	B	C	D	E	F
1	Power-One Sample Z test					
2	Two-tails					
3						
4	To find POWER				To find SAMPLE SIZE	
5	Ho Mean	100			Ho Mean	100
6	H1 Mean	100.5			H1 Mean	100.5
7	SD (Sigma)	5			SD (Sigma)	5
8	Alpha	0.05			Alpha	0.05
9	Sample Size	6			Power	0.9
10						
11	Xbar critical 1	104.00076				
12	Xbar critical 2	95.99924				
13	Effect size	0.1			Effect size	0.1
14						
15	Beta	0.9568288			Sample Size	1051
16	Power	0.0431712				
17						

Figure 11.4.

as before, we would like to know the sample size required to obtain a power of 0.90. We enter all the required information in the SAMPLE SIZE template and we found that with a sample of 1,051 observations, we have a high probability to find a significant effect if there is actually one.

Power computations are substantially more complex for other statistical tests, and specialized shareware software is available (http://www.psychologie.hhu.de/arbeitsgruppen/allgemeine-psychologie-und-arbeitspsychologie/gpower.html).

WORKED EXAMPLE 11.4

In the fourth worksheet of the Excel file for this chapter, we have two empty templates, one for the Z-test and the other for finding power for the Z-test. We will use them for the following analysis (see Figure 11.5):

The "consideration of future consequences" (CFC) as a personality trait is defined by the extent to which individuals consider the potential future consequences of their current behaviors and the extent to which they are influenced by those imagined consequences. Individuals, who score high on a CFC inventory, typically focus on the future implications of their behavior. Those scoring low on CFC typically focus more on their immediate needs and concerns.

In a large US college population, a research team applied a 12 Likert-item CFC scale (scores from 1 = extremely uncharacteristic to 5 = extremely characteristic). The reported average in the scale was $\mu = 3.51$ with a standard deviation of $\sigma = 0.61$. Suppose that in a sample of $N = 100$ college students from a recently hurricane-affected area, we find a mean of $\bar{X} = 3.7$ in this CFC scale.

1. Use the Z-test template to test the null hypothesis H_0: $\mu = 3.51$ against a two-tail alternative. Use an alpha level of .05. Do you reject or retain the null hypothesis?

Yes, reject the null hypothesis. The sample does not seem to be a random sample from the large US college population.

	A	B	C	D	E	F
1	Z test for Hypothesis about a single mean				Power-One Sample Z test	
2					Two-tails	
3	Null Hypothesis H0	3.51				
4	Alternative	<>	(<>, <, >)		To find POWER	
5					Ho Mean	3.51
6	Standard Deviation	0.61			H1 Mean	3.7
7					SD (Sigma)	0.61
8	Sample Mean	3.7			Alpha	0.05
9	Sample Size	100			Sample Size	100
10						
11	Standard Error	0.061			Xbar critical 1	3.6295578
12	Zobt	3.114754			Xbar critical 2	3.3904422
13	p-value	0.001841			Effect size	0.3114754
14						
15	Effect size	0.311475			Beta	0.1240882
16					Power	0.8759118
17	95% C.I.	3.58044	3.81956			
18						

Figure 11.5.

2. Is the effect size of the test large?
   It is between a small and medium effect size.

3. If we run another test with the null hypothesis H_0: $\mu = 3.65$, would you reject the null hypothesis? Why yes or not?
   No, I will not reject the null hypothesis, because the value of 3.65 is inside the 95% CI for the mean.

4. Calculate the power of the test (use as specific alternative the sample mean of 3.7). Is this a large power?
   Power of 0.8759119, yes it is a large power.

The final output for the templates is in worksheet 5 of the Excel file for this chapter.

12 One-Sample t-Test and CI

In the previous chapter, we introduced the statistical hypothesis testing framework by presenting the Z-test for one sample mean. However, the Z-test for one sample made the assumption that we know the population standard deviation σ. However, in many settings, we do not know σ. In these circumstances, the only information available is the sample standard deviation, S. We can use S as an estimator of the population standard deviation in the test-statistic formula for the Z_{obt}. Nevertheless, the test-statistic receives now as new label t_{obt} because we will use as sampling distribution a new probability distribution called the ***t-distribution***

$$t_{obt} = \frac{\overline{X} - \mu}{S/\sqrt{N}}$$

The t-distribution, as the standard normal distribution, is a symmetric distribution with mean zero and positive and negative values. However, the shape of the t-distribution depends on its ***degrees of freedom***, which for the one-sample t-test is the sample size minus one, i.e., $N - 1$. Furthermore, the larger the sample size, and, therefore, the larger the degrees of freedom, the closer the t-distribution resembles the standard normal distribution.

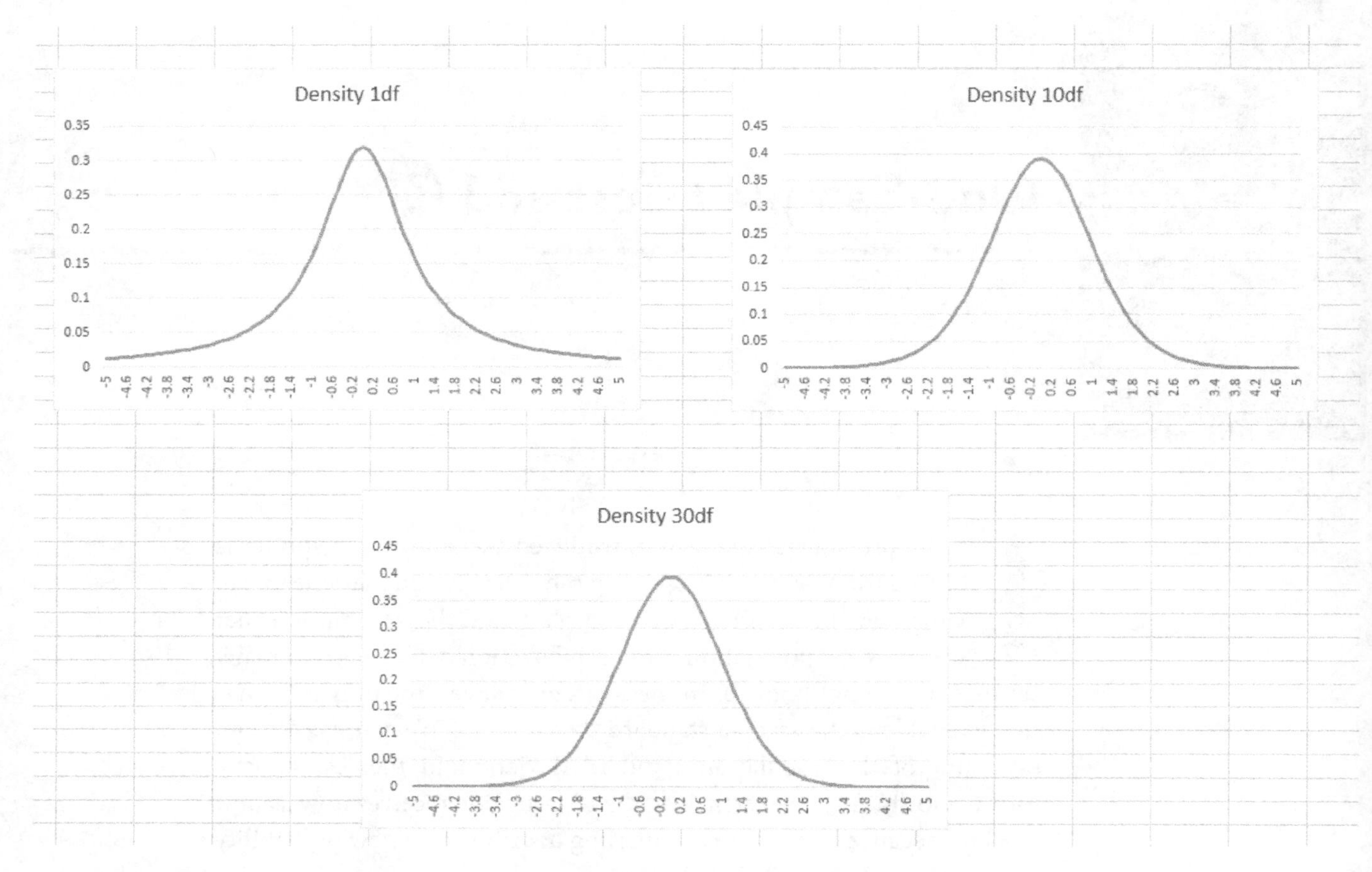

Figure 12.1.

In the first worksheet of the Excel file for this chapter, we have the graphs of three t-distributions (see Figure 12.1): (a) t with 1 degree of freedom, (b) t with 10 degrees of freedom, and (c) t with 30 degrees of freedom. With only 1 degree of freedom, the t-distribution has long "heavy" or "thick" tails, which make values of −5 or +5 not unlikely to occur. The larger the degree of freedom, the closer the distribution is to the standard normal. With 30 degrees of freedom, the t-distribution has "light" or "thinner" tails that, as in a normal distribution, make values beyond +3 or −3 to be very unlikely.

As in the case of the Z-test for one sample, hypothesis testing for one sample mean using a t-distribution can be thought as "assessing the membership" of the sample mean in a sampling distribution defined by the value of the population mean stated in the null hypothesis. For example, we would like to test some hypothesis about the average number of hours that college students watch regular cable TV:

> Nielsen, a media research company, reports that individuals in the range of 18 to 34 years watch traditional cable TV on average 13 hours per week. In a sample of $N = 45$ college students, we found that the average number of hours watching cable TV per week is $\bar{X} = 6.3$ hours with standard deviation $S = 6.8$. How likely is it to obtain this sample mean, or a more extreme one, if our sample is a random sample from the general 18- to 34-year-old population in the United States?

The ***statistical hypotheses*** are: H_0: $\mu = 13$ vs. H_1: $\mu \neq 13$

The ***test-statistic*** value is: $t_{obt} = \dfrac{6.3 - 13}{6.8/\sqrt{45}} = -5.623$

The ***sampling distribution*** is: t-distribution with $45 - 1 = 44$ degrees of freedom.

The ***p-value*** is: Using the t-distribution:

$$p-value = P(t_{44} \leq -5.623) + P(t_{44} \geq 5.623) = 0.000001$$

Using ***critical value α*** = .05: Because p-value < α, reject the null hypothesis. The college group does not come from the general 18–34 population

The ***effect size***: $d = \dfrac{|6.3 - 13|}{6.8} = 0.9853$, a large effect size.

OBJECTIVES

12.1

1. To compute a t-test for one sample using Excel basic functions.
2. To use the chapter template to compute a t-test for one sample, its effect size, and 95% confidence interval (CI) for the mean.

t-TEST FOR ONE SAMPLE

12.2

Continuing with the TV watching behavior of college students, we would like to see if students' majors play a role in the average number of hours they watch cable TV.

Suppose that we obtain two small samples of size $N = 6$: one of Psychology majors and the other one of students from other majors.

In the second worksheet in the Excel file for this chapter, we have the data, the mean, standard deviations, and sample size for each of the samples (see Figure 12.2).

- To obtain the PSYC mean, we put in B8 =AVERAGE(B2:B7). We can drag this function to cell C8 to obtain the mean for OTHER majors.
- For PSYC standard deviation, we enter in B9 =STDEV.S(B2:B7). Again, we drag the function to C9 to obtain the standard deviation for OTHER.
- For PSYC sample size, we enter in B10 =COUNT(B2:B7). Drag the cell to B11 to obtain the OTHER sample size.

	A	B	C
1		PSYC	OTHER
2		4	4
3		13	20
4		2	5
5		6	28
6		14	4
7		5	20
8	Mean	7.333333333	13.5
9	SD	4.966554809	10.46422477
10	N	6	6
11			
12	t obt	-2.794782784	0.117041147
13	p-value	0.038228317	0.911383447
14			
15	Effect Size	1.34231211	0.047781848
16			
17	Half-width	5.212079624	10.98153042
18	Lower 95%	2.12125371	2.518469584
19	Upper 95%	12.54541296	24.48153042
20			

Figure 12.2.

We want to test the null hypothesis that the samples come from a population with mean 13, vs. a two-tail alternative of coming from a population different than 14.

$$H_0: \mu = 13 \qquad \text{vs.} \qquad H_1: \mu \neq 13$$

Test Statistic: t_{obt}

We compute t_{obt} using $t_{obt} = \frac{\bar{X} - \mu}{S/\sqrt{N}}$.

- In cell B12, we enter the equation =(B8-13)/(B9/SQRT(B10)) to get the t_{obt} for the PSYC sample.
- We drag the equation to cell C12 in order to get the t_{obt} for the OTHER sample.

p-Value for Two Tails

To obtain the p-value for the t_{obt}, we consider the **degrees of freedom** and the **alternative hypothesis**. In this example, degrees of freedom is $N - 1 = 6 - 1 = 5$ for each sample. The alternative hypothesis is "*not-directional*" or "*two tails*." Thus, the probability of "more extreme values" for t_{obt} has to look for ***both*** tails of the t-distribution with 5 degrees of freedom. Because the t-distribution is symmetric, we can use the absolute value of t_{obt}, find the area above that value, and multiply it by two, i.e.,

$$p = 2 \times P(t_{N-1} > abs(t_{obt}))$$

Fortunately, the Excel function T.DIST.2T([value], [degrees of freedom]) returns directly the two-tail p-value. This function requires as input the **absolute value** of t_{obt} and the degrees of freedom.

- In cell B13, we enter =T.DIST.2T(ABS(B12), B10-1) to obtain the p-value for the PSYC sample.
- Drag the formula to cell C13 to obtain the p-value for the OTHER sample.

Thus, using an alpha level of .05, we will reject the null hypothesis that the "PSYC" sample comes from a population with mean 13 hours. On the other hand, we will retain the null hypothesis for the "OTHER" sample.

Effect Size

The Cohen's d effect size is a measure of the importance of the departure from the null hypothesis. It expresses the difference between the sample mean and the hypothesis, but now in terms of the standard deviation of the sample, i.e., $d = \frac{|\bar{X} - \mu|}{S}$.

- We enter in cell B15, the expression =ABS(B8-14)/B9 to obtain the effect size for the PSYC sample.
- Drag the formula to cell C15 to obtain the effect size for the OTHER sample.

The effect size is quite large for the Psychology sample but very small for the "Other" sample.

Confidence Interval

When we do not know the population standard deviation, we can use the t-distribution to obtain the 100(1 – α)% CI for the population mean. The formula is given by:

$$\bar{X} \pm t_{a/2;N-1}\left(S/\sqrt{N}\right)$$

where $t_{a/2;N-1}$ is the value of t from a t-distribution with N – 1 degrees of freedom that defines the lower α/2 area and the upper α/2 area. Excel has the function CONFIDENCE.T (alpha, Std.Dev, N) that produces the value $t_{a/2;n-1}(S/\sqrt{N})$ once we provide the desired alpha, the sample standard deviation, and the sample size. This value is equal to the half-width of the CI. We will compute the 95% CI for the means.

- We enter in B17 the function =CONFIDENCE.T(.05, B9, B10) to obtain the half-width of the interval for the PSYC sample.
- Drag the function to cell C17 to obtain the interval half-width for the OTHER sample.
- In cell B18, we enter = B8-B17 to obtain the lower limit for the 95% CI for the PSYC sample.
- In cell B19, we enter = B8 + B17 to get the upper limit for the 95% CI for the PSYC sample.
- For the OTHER sample, drag the equation in B18 to C18 and the equation in B19 to C19.

The value of 134 is not inside the CI for the Psychology majors, but this value is inside the "Other" majors.

A Template for t-Test for One Sample

In the third worksheet of the Excel file for this chapter, we have an Excel template to perform the one-sample t-test. The template obtains the t_{obt}, p-value, effect size, and the 95% CI for the population mean. In the gray cells we enter as input: (1) the value of the null hypothesis, (2) the direction of the alternative hypothesis, <> for a two-tail test, < for a one-tail lower test, or > for a one-tail upper test, (3) the sample standard deviation, (4) the sample mean, and (5) the sample size. Below we use the template to repeat the t-tests for the PSYC (see Figure 12.3) and OTHER major (see Figure 12.4) groups.

	A	B	C
1	t test for Hypothesis about a single mean		
2			
3	Null Hypothesis H0	13	
4	Alternative	<>	(<>, <, >)
5			
6	Sample Standard Deviation	4.966555	
7			
8	Sample Mean	7.333333	
9	Sample Size	6	
10			
11	Degrees of freedom	5	
12	Standard Error	2.027588	
13	t-obt	-2.79478	
14	p-value	0.038228	
15			
16	Effect size	1.140965	
17			
18	95% C.I.	2.121254	12.54541
19			
20			

Figure 12.3.

	A	B	C
1	t test for Hypothesis about a single mean		
2			
3	Null Hypothesis H0	13	
4	Alternative	<>	(<>, <, >)
5			
6	Sample Standard Deviation	10.46422	
7			
8	Sample Mean	13.5	
9	Sample Size	6	
10			
11	Degrees of freedom	5	
12	Standard Error	4.272	
13	t-obt	0.117041	
14	p-value	0.911383	
15			
16	Effect size	0.047782	
17			
18	95% C.I.	2.518475	24.48153
19			
20			

Figure 12.4.

12.3 WORKED EXAMPLE

Several articles in shopping magazines mention that women over 18 years old own an average of 20 pairs of shoes. As an exercise in an Apparel and Merchandise class, the instructor asked a sample of $N = 94$ female students how many pairs of shoes, of any type, they own. The data and the template for one-sample *t*-test are in the fourth worksheet in the Excel file for the present chapter. We would like to test the hypothesis that the sample comes from a population with mean 20 (see Figure 12.5).

	A	B	C	D	E
1	Pairs				
2	20		MEAN	31.0851064	
3	35		SD	22.7825908	
4	33		N	94	
5	19				
6	15				
7	21				
8	20		t test for Hypothesis about a single mean		
9	15				
10	20		Null Hypothesis H0	20	
11	40		Alternative	<>	(<>, <, >)
12	38				
13	20		Sample Standard Deviatio	22.7825908	
14	25				
15	40		Sample Mean	31.0851064	
16	9		Sample Size	94	
17	20				
18	30		Degrees of freedom	93	
19	20		Standard Error	2.34984482	
20	30		t-obt	4.71737806	
21	20		p-value	8.3904E-06	
22	17				
23	12		Effect size	0.4865604	
24	36				
25	15		95% C.I.	26.4187803	35.7514325
26	25				
27	16				

Figure 12.5.

1. First, find and report the sample average number of pairs of shoes, the standard deviation, and the number of observations (round mean and SD to two decimals).
   - In cell D2, enter =AVERAGE(A2:A95)
   - In cell D3, enter =STDEV.S(A2:A95)
   - In cell D4, enter = COUNT(A2:A95)

   Mean = 31.09 SD = 22.78 $n = 94$

2. Use the *t*-test template to test the two-tail null hypothesis $\mu = 20$. State your decision about the null hypothesis, i.e., do you retain it or reject it. Use as alpha .05
   - Enter 20 as the value for null hypothesis.
   - Type <> as alternative hypothesis
   - Enter o copy the sample standard deviation, sample mean, and sample size.

   We reject the null hypothesis because the *p*-value (which it is in scientific notation) of .0000083904 is smaller than .05

3. If you reject the null hypothesis, report your 95% CI to estimate the population average number of pairs (again, round the values to two decimals). Is the sample coming from a population with larger or smaller mean than stated in the null hypothesis?

   The 95% CI for the mean is between 26.42 and 35.75 pairs of shoes.

   The sample seems to come from a population with a larger mean than 20, because the CI does not include the value 20. In fact, the CI includes values substantially larger than 20.

The final output is in worksheet 5 of the Excel file for this chapter.

13 t-Test for Paired Samples and CI

Statistical hypothesis testing is the standard analysis for research with randomized ***experimental designs***. Experimental design refers to many aspects of a study, being one of them the way the experimenter allocates participants to different treatments or conditions. The allocation to one condition follows some random procedure, such as tossing a coin or a die, to assign participants to one of the research groups. Other times, the experimenter assigns participants to more than one condition, in which case the order in which participants work under each condition is determined at random. In this chapter, we introduce the ***two paired samples or two correlated samples design***. In this design, the participants are observed under two experimental conditions. The purpose of the design is to know if the participants' performance scores differ under the two conditions. In this design, the experimenter will obtain each participant's difference between both conditions score, D, and will use the t-test for one sample on the difference scores. We refer to this test as the ***paired samples t-test*** or ***paired t-test***.

For each participant we have two scores: the participant's performance in condition 1, X_1, and the same participant's performance in condition 2, X_2. The difference between these two scores, $D = X_1 - X_2$, becomes the input for assessing any differential effect of the two conditions. Because most of the times, we want to test for the equality of means under the two conditions, the null hypothesis for the paired t-test is that the average D

comes from a sampling distribution with population mean 0. Therefore, the two-tail statistical hypotheses for the test are:

$$H_0: \quad \mu_D = 0. \qquad H_1: \mu_D \neq 0.$$

The test-statistic, simplifies to:

$$t_{obt} = \frac{\bar{D}}{S_D/\sqrt{N}}$$

As in the previous case for the t-test for one sample, the sampling distribution is a t-distribution with $N - 1$ degrees of freedom. The p-value for the test-statistic assesses how likely is to obtain the sample mean difference, or a more extreme value, if the sample mean actually comes from a sampling distribution with mean zero. An example follows:

> In a wine tasting, $N = 20$ judges are asked to rate two red wines using a five-star scale, i.e., a superb top-tier wine will receive five stars and a worthless wine will receive only one star. The order in which the judges drink the wines is at random, i.e., some judges drink Wine1 first, while others will drink Wine2 first. For each judge, the difference between the ratings is computed as $D = \text{Wine1} - \text{Wine2}$. The average difference for the judges was $\bar{D} = 0.95$ and the standard deviation was $S_D = 0.88$. How likely is to obtain this sample mean difference, or a more extreme difference, if our sample is a random sample from a population with mean difference 0, i.e., if the two wines are considered of equal quality?

The ***statistical hypotheses*** are: $H_0: \mu = 0$ vs. $H_1: \mu \neq 0$

The ***test-statistic*** value is: $t_{obt} = \frac{0.95}{0.88/\sqrt{20}} = 4.828$

The ***sampling distribution*** is: t-distribution with $20 - 1 = 19$ degrees of freedom.

The ***p-value*** is: Using the t-distribution:

$$p-value = P(t_{19} \leq -4.828) + P(t_{19} \geq 4.828) = 0.000117$$

Using ***critical value α*** =.05: Because p-value < α, reject the null hypothesis, wines are different in quality.

The ***effect size***: $d = \frac{|0.95|}{0.88} = 1.079$, a large effect size.

OBJECTIVES

13.1

1. To computer a t-test for paired samples using Excel basic functions.
2. To use the Data ToolPak to compute a t-test for paired samples.
3. To use the chapter template to compute a t-test for paired samples, its effect size, and 95% confidence interval (CI) for the mean difference.

t-TEST FOR PAIRED SAMPLES USING EXCEL BASIC FUNCTIONS

13.2

A college math professor suspects her freshman students come to her course without a good high-school algebra background. Nevertheless, she feels that her students catch up with their algebra knowledge by the end of the semester. The professor selects a sample of ten students to take the same algebra test at the beginning (ALGEBRA1) and at the end (ALGEBRA2) of the semester. Although she expects the students to improve their algebra scores, she decides to run a two-tail test, i.e., using as alternative hypothesis that the mean in the two testing occasions will *differ* but without stating which one will be larger or smaller than the other. She wants to test the hypothesis using α = .05, and she defines the difference as D = ALGEBRA2 – ALGEBRA1 (see Figure 13.1).

In the first worksheet in Excel file for this chapter, we have the algebra scores at the beginning and at the end of the semester. We perform the following computations:

- To obtain the difference scores D, enter in D2 the function =C2-B2. Drag this equation to the remaining rows.
- Finding the average of the differences: Enter in cell D12 =AVERAGE(D2:D11)
- Finding the standard deviation of the differences: Enter in cell D13 = STDEV.S(D2:D11)
- Finding the number of observations: In cell D14, enter =COUNT(D2:D11)

	A	B	C	D
1		Algebra 1	Algebra 2	D
2		13	20	7
3		18	26	8
4		10	16	6
5		25	31	6
6		19	27	8
7		14	17	3
8		19	27	8
9		18	23	5
10		19	23	4
11		16	24	8
12	Mean			6.3
13	SD			1.82878223
14	N			10
15				
16	tobt			10.89377889
17	p-value			1.74712E-06
18	Effect Size			3.444915363
19	HalfWidth			1.308231998
20	Lower 95%			4.991768002
21	Upper 95%			7.608231998
22				
23	Using t.TEST			1.74712E-06
24				

Figure 13.1.

We want to test the null hypothesis that the sample average difference comes from a population with mean 0 vs. a two-tail alternative hypothesis, i.e.,

$$H_0 : \mu_D = 0 \quad H_1 : \mu_D \neq 0$$

Test Statistic: t_{obt}

We compute t_{obt} using $t_{obt} = \frac{\bar{D}}{S_D / \sqrt{N}}$. Where $\bar{D}$ is the average of the difference scores, S_D is the standard deviation of the difference scores, and N is the sample size.

- In cell D16, we enter the equation =D12/(D13/SQRT(D14)) to get the t_{obt}.

p-Value for Two Tails

As in the case of the *t*-test for one sample, the Excel function T.DIST.2T (tvalue, df) returns directly the two-tail *p*-value. This function requires as input the **absolute value** of t_{obt} and the degrees of freedom. In our example:

- In cell D17, we enter =T.DIST.2T(ABS(D16), D14-1) to obtain the *p*-value.

Thus, for a *p*-value of .00000174712 and an alpha level of .05, we will reject the null hypothesis because *p*-value < alpha. The average *D* comes from a sampling distribution with mean different from zero. Because the average *D* is larger than zero, we conclude that the average score in Algebra 2 was significantly higher than the average score in Algebra 1. The professor was right, her students improved their algebra knowledge.

The t-TEST Function

Excel function T.TEST produces directly the *p*-value for one- or two-tails *t*-tests once we enter the range of the two sets of scores, the number of tails of the alternative hypothesis, and the type of *t*-test. The form of the function is T.TEST(range1, range2, #tails, type). The last entry in the function, the type, can take values of:

1 for *t*-test with paired groups,
2 for *t*-test with two independent groups with equal variance, and
3 for *t*-test with two independent groups with unequal variances.

In our case, we insert the function =T.TEST(B2:B11, B2:C11, 2, 1) in cell D23. The number of tails is 2 and the type of test is 1. The resulting value is the same as the one we found in cell D17.

Effect Size

The Cohen's *d* effect size in the case of the paired *t*-test is simply $d = \frac{|\bar{D}|}{S_D}$. It expresses the size of the difference in terms of the standard deviation of the differences.

- In our example, we enter in cell D18, the expression =ABS(D12)/D13. The effect size of 3.449 is large.

Confidence Interval

Similarly to the case of the t-test for one sample, the $100(1 - \alpha)\%$ CI for the population mean difference can be written as:

$$\overline{D} \pm t_{\alpha/2;N-1}\left(S_D/\sqrt{N}\right)$$

where $t_{\alpha/2;n-1}$ is the value of t from a t-distribution with $N - 1$ degrees of freedom that defines the lower $\alpha/2$ area and the upper $\alpha/2$ area. We use the Excel function CONFIDENCE.T (alpha, Std.Dev, N) to obtain the value $t_{\alpha/2;N-1}(S_D/\sqrt{N})$. This function requires as input the desired alpha, the sample standard deviation of the differences, and the sample size. This value is equal to the half-width of the CI.

- We enter in D19 the expression =CONFIDENCE.T(.05, D13, D14) to obtain the half-width of the interval.
- For the lower limit of the CI, we enter in D20 the expression =D12-D19
- For the upper limit of the CI, we enter in D21 the expression =D12+D19

Thus, possible values for the population difference are between 4.99 and 7.61.

Using the Template for One-Sample t-Test

We can apply our t-test for one-sample template to the case of a t-test for paired samples. In the second worksheet of the Excel file for this chapter, we have the template and the data of the algebra scores example, the difference scores D, and the mean and standard deviation for D (see Figure 13.2).

In the gray cells, we enter (1) the value of the null hypothesis (most of the time zero), (2) the direction of the alternative hypothesis (<> for a two-tail test, < for a one-tail lower test, or > for a one-tail upper test), (3) the sample standard deviation for D, (4) the sample mean, for D, and (5) the sample size.

We reject the null hypothesis because p-value = .00000175 is less than .05. The effect size of 3.449 is large. Also, the 95% CI states that the difference between the two algebra tests can be between 4.99 and 7.61 points.

	A	B	C	D	E	F	G	H
1		Algebra 1	Algebra 2	D		t test for Hypothesis about a single mean		
2		13	20	7				
3		18	26	8		Null Hypothesis H0	0	
4		10	16	6		Alternative	<>	(<>, <, >)
5		25	31	6				
6		19	27	8		Sample Standard Deviatio	1.828782	
7		14	17	3				
8		19	27	8		Sample Mean	6.3	
9		18	23	5		Sample Size	10	
10		19	23	4				
11		16	24	8		Degrees of freedom	9	
12	Mean			6.3		Standard Error	0.578312	
13	SD			1.828782		t-obt	10.89378	
14	N			10		p-value	1.75E-06	
15								
16						Effect size	3.444915	
17								
18						95% C.I.	4.991768	7.608232
19								

Figure 13.2.

USING DATA ANALYSIS TOOLPAK FOR *t*-TEST FOR PAIRED SAMPLES 13.3

The Data Analysis ToolPak includes an option for running a *t*-test for paired samples. In the third worksheet of the Excel file for this chapter, we copy the data for the algebra scores and we use the Data Analysis ToolPak (see Figure 13.3).

- Click on the Data tab in the top Excel menu, and then click on the Data Analysis icon.
- Select the option "t-Test: Paired Two Sample for Means" and press OK.
- In the Input Variable Range 1, enter the range for the Algebra 2 scores (C2:C11).
- In the Input Variable Range 2, enter the range for the Algebra 1 scores (B2:B11).
- Type in 0 as the hypothesized mean difference.
- Leave the alpha value at .05.
- Click on "Output Range" and select as first cell for the table E1. Then click OK

	A	B	C	D	E	F	G
1		Algebra 1	Algebra 2		t-Test: Paired Two Sample for Means		
2		13	20				
3		18	26			*Variable 1*	*Variable 2*
4		10	16		Mean	23.4	17.1
5		25	31		Variance	22.0444444	16.9888889
6		19	27		Observations	10	10
7		14	17		Pearson Correlation	0.92208502	
8		19	27		Hypothesized Mean Differenc	0	
9		18	23		df	9	
10		19	23		t Stat	10.8937789	
11		16	24		P(T<=t) one-tail	8.7356E-07	
12					t Critical one-tail	1.83311293	
13					P(T<=t) two-tail	1.7471E-06	
14					t Critical two-tail	2.26215716	
15							

Figure 13.3.

The output provides the mean, variance (no standard deviation), and sample size for each of the two conditions. In addition, we have the value of the null hypothesis (0 in our case), the value of the test-statistic "*t* Stat" or t_{obt}, and the *p*-values for the one-tail and two-tail tests. The correlation between the two set of scores is displayed and labeled "Pearson Correlation."

13.4 WORKED EXAMPLE

A consumer researcher runs a blind test comparing two brands of potato chips: brand A and brand B. Twenty participants taste and rate both brands on a scale from 1 = "I extremely dislike it" to 10 = "I extremely like it." The researcher wants to know if there is any significant difference between the two brands of chips. She decides to compute the difference as D = brand B – brand A, and test if the average difference can be considered as coming from a sampling distribution with a difference mean zero vs. different from zero. The data and an empty template are in the fourth worksheet in the lab file (see Figure 13.4).

	A	B	C	D
1		Brand A	Brand B	D
2		9	4	-5
3		3	7	4
4		2	6	4
5		6	8	2
6		5	7	2
7		7	7	0
8		8	8	0
9		3	6	3
10		9	7	-2
11		3	8	5
12		5	9	4
13		2	8	6
14		9	7	-2
15		6	4	-2
16		2	6	4
17		5	7	2
18		7	6	-1
19		2	5	3
20		6	5	-1
21		3	6	3
22	Mean	5.1	6.55	1.45
23	SD	2.511028	1.356272	2.910507
24	N	20	20	20
25	p-value	0.03816		
26				

F	G	H
t test for Hypothesis about a single mean		
Null Hypothesis H0	0	
Alternative	<>	(<>, <, >)
Sample Standard Deviation	2.910507	
Sample Mean	1.45	
Sample Size	20	
Degrees of freedom	19	
Standard Error	0.650809	
t-obt	2.227996	
p-value	0.03816	
Effect size	0.498195	
95% C.I.	0.087841	2.812159

J	K	L
t-Test: Paired Two Sample for Means		
	Brand A	*Brand B*
Mean	5.1	6.55
Variance	6.305263158	1.839473684
Observations	20	20
Pearson Correlation	-0.047908182	
Hypothesized Mean Difference	0	
df	19	
t Stat	-2.227995504	
P(T<=t) one-tail	0.019079943	
t Critical one-tail	1.729132812	
P(T<=t) two-tail	0.038159886	
t Critical two-tail	2.093024054	

Figure 13.4.

1. Obtain the difference scores between potato chips brands. Compute and report the mean, standard deviation, and sample size for the differences (round the results to two decimals).
   - In cell D2, enter =C2-B2. Drag the formula to the remaining cells.
   - For mean D, in cell D22 enter =AVERAGE(D2:D21)
   - In cell D23, enter =STDEV.S(D2:D21) to obtain the standard deviation.
   - For number of observations, in cell D24, enter =COUNT(D2:D21)

   The mean difference $\bar{D}$ is 1.45 and standard deviation of the difference S_D is 2.91, with $N = 20$.

2. Using the T.TEST function to obtain the p-value
   - In cell B25, enter =T.TEST(B1:B21, C1:C21, 2, 1). (Notice that B1 and C1 contain the headers, which is ok.) For alpha of .05 we will reject the null hypothesis.

3. Use the template for the one sample *t*-test to test the null hypothesis of mean difference 0 (vs. a two-tail alternative). For an alpha of .05, state your conclusion about the null hypothesis.
   - Enter as null hypothesis 0, as type of alternative <>, the values for the standard deviation of D, the average D, and the sample size.

For alpha .05, we reject the null hypothesis that the sample comes from a population with mean zero.

4. Report the 95% CI for the mean difference (round up to two decimals). Would you say that brand B is more liked than brand A? Would you say that the average difference comes from a population with mean 3? From a population with mean difference of 2?

The 95% CI for the difference is (0.09, 2.81). No the average difference does not come from a population with mean difference 3. Yes, the average difference may come from a population with mean difference 2.

5. Repeat the analysis using the Data ToolPak, using also the headers for the brands.
   - Click on the Data tab in the top Excel menu, and then click on the Data Analysis icon.
   - Select the option "t-Test: Paired Two Sample for Means" and press OK.
   - In the Input Variable Range 1, enter the range for brand A, including the header (B1:B21).
   - In the Input Variable Range 2, enter the range for brand B, including the header (C1:C21).
   - Type in 0 as the hypothesized mean difference.
   - Check the box "Labels."
   - Leave the alpha value at .05.
   - Select "Output Range" and select as first cell for the table J1. Then click OK.

The final output of the analysis is in worksheet 5 in the Excel file for this chapter.

14 t-Test for Two Independent Groups (Equal Variances) and Confidence Interval (CI)

The experimental design that represents the basic idea of experimentation in the mind of the general public is the ***two independent groups design***. In this design, participants are assigned at random to only one of two experimental conditions. The ***random assignment***, a key component of this design, aims to obtain two groups of participants as similar as possible before the actual application of the experimental conditions. Therefore, if the two groups differ in their responses after the experimental conditions are applied, we can consider the difference as the product of the experimental treatments.

In this design, the word "independent" refers to the fact that different groups of participants are assigned to each of the experimental conditions. Therefore, in contrast to the paired sample design, the comparison between the two groups is done at the level of the group's mean response, i.e., we compare the average $\overline{X}_1$ of group 1 vs. $\overline{X}_2$ of group 2 by using their difference $\overline{X}_1 - \overline{X}_2$. We analyze the difference between the two means with a ***t-test for two independent groups***. Because most of the time we want to test for the equality of the two means, the null hypothesis for the two independent groups' t-test is that the difference between the two samples means comes from a sampling distribution of difference between means with a mean 0, or, for the two-tail alternative hypothesis:

$$H_0: \mu_1 - \mu_2 = 0 \qquad H_1: \mu_1 - \mu_2 \neq 0$$

One version of the t-test for two independent groups assumes that the populations from where the groups come from have ***equal variances*** or standard deviations. Under this assumption, the test-statistic uses a ***pooled or weighted standard deviation, S_W***, to obtain the standard error of the difference between two means. This S_W is a weighted average of the standard deviations of each group. Thus, this version of the t-test is referred as ***t-test for two independent groups with equal variances.***

$$t_{obt} = \frac{\bar{X}_1 - \bar{X}_2}{S_W \sqrt{\frac{1}{N_1} + \frac{1}{N_2}}}$$

The sample size for each group is represented by N_1 and N_2. The test-statistics follows a t-distribution with degrees of freedom of $(N_1 - 1) + (N_2 - 1)$ or $N_1 + N_2 - 2$.

If we assume that the two groups come from populations with ***unequal variances***, the t-test uses directly the standard deviations of the groups, and as degrees of freedom (that can take decimal values) a function of these standard deviations. We will not cover this ***t-test for two independent groups with unequal variances.***

As in the previous t-tests, our testing procedure assesses how likely the difference between the two sample means can be considered a member of a sampling distribution of mean differences with mean zero. For example:

> A publisher needs to know if **bold** or *italic* fonts better help students to remember key concepts in a textbook chapter. She prints two versions of the same chapter, one using **bold** and the other using *italics* to highlight key concepts. She recruits 16 volunteers and assign at random an equal number to read the **bold** or the *italic* version of the chapter. Afterwards, she tests the students using a 10-item quiz. For the ***bold*** version, the average in the quiz is $\bar{X}_1 = 7.75$ with a standard deviation of $S_1 = 1.25$. For the *italic* version, the quiz average is $\bar{X}_2 = 6.25$ with a standard deviation of $S_2 = 1.40$. How likely is it to obtain a difference between means of 1.5 (7.75 – 6.25) if our sample is a random sample from a population with the difference between means equal to 0, i.e., if the two fonts produce equal recall?

The ***statistical hypotheses*** are: $H_0\text{: } \mu_1 - \mu_2 = 0$ vs. $H_1\text{: } \mu_1 - \mu_2 \neq 0$

The ***pooled standard deviation*** is: $S_W = \sqrt{\frac{(N_1 - 1)S_1^2 + (N_2 - 1)S_2^2}{N_1 + N_2 - 2}}$

$$= \sqrt{\frac{(8-1)1.25^2 + (8-1)1.40^2}{8+8-2}} = 1.327$$

The ***test-statistic*** value is: $t_{obt} = \frac{\bar{X}_1 - \bar{X}_2}{S_W\sqrt{1/N_1 + 1/N_2}} = \frac{7.75 - 6.25}{1.327\sqrt{1/8 + 1/8}} = 2.26$

The ***sampling distribution*** is: t-distribution with 16 – 2 = 14 degrees of freedom.

The ***p-value*** is: Using the t-distribution:

$$p - value = P(t_{14} \leq -2.26) + P(t_{14} \geq 2.26) = 0.040$$

Using ***critical value α*** = .05: Because p-value < α, reject the null hypothesis, the fonts differ.

The ***effect size***: $d = \frac{|\bar{X}_1 - \bar{X}_2|}{S_W} = \frac{|7.75 - 6.25|}{1.327} = 1.13$ a large effect size.

OBJECTIVES

14.1

1. To compute a t-test for two independent samples with equal variance samples using Excel basic functions.
2. To use the Data ToolPak to compute a t-test for two independent samples with equal variances.
3. To use the chapter template to compute a t-test for two independent samples with equal variance, its effect size, and 95% confidence interval (CI) for the mean difference.

t-TEST FOR TWO INDEPENDENT GROUPS

14.2

In a reading comprehension test, examinees read a passage and answer a set of multiple-choice items about the passage. Some researchers argue that examinees can answer correctly some of the items even when the passage is not provided to them. In a study, 16 students were assigned at random to two groups. The "Without" group was presented with five multiple-choice reading comprehension items but no passage. The "With" group read the passage and the five multiple-choice items. Unfortunately, the data for two of the participants in the "Without" group was lost.

	A	B	C
1		Without	With
2		2	4
3		1	3
4		1	3
5		3	4
6		2	3
7		3	4
8			2
9			5
10	Mean	2	3.5
11	SD	0.894427191	0.9258201
12	N	6	8
13			
14	df	12	
15	Pooled SD	0.912870929	
16	Standard Error	0.493006649	
17	tobt	-3.042555317	
18	p-value	0.010225204	
19			
20	Effect Size	1.643167673	
21			
22	HalfWidth	1.074169211	
23	Lower 95%	-2.574169211	
24	Upper 95%	-0.425830789	
25			
26	Using t.TEST	0.010225204	
27			

Figure 14.1.

The data is in the first worksheet of the Excel file for this chapter. First, we obtain the means, standard deviations, and counts for each group (see Figure 14.1).

- Finding the averages: For the "Without" group, enter in cell B10 =AVERAGE(B2:B9). Notice that cells B8 and B9, although empty, are included in the range. This is done because when dragging the function in B10 to cell C10, in order to obtain the mean of the "With" group, the range will include the values in cells C8 and C9.

- Finding the standard deviation: Enter in cell B11 =STDEV.S(B2:B9). Drag the function to cell C11.
- Finding the number of observations: In cell B12, enter =COUNT(B2:B9). Drag the function to cell C12.

We want to test the null hypothesis that the samples come from populations with the same mean versus the alternative hypothesis that they come from populations with *different* means. We assume equal population variances.

$$H_0 : \mu_1 - \mu_2 = 0 \quad H_1 : \mu_1 - \mu_2 \neq 0$$

Test Statistic: t_{obt}

The degrees of freedom for the test is $df = N_1 + N_2 - 2$:

- Enter in cell B14 =B12+C12-2.

  For the pooled standard deviation, $S_W = \sqrt{\dfrac{(N_1 - 1)S_1^2 + (N_2 - 1)S_{21}^2 +}{df}}$:

- Enter in cell B15 =SQRT(((B12-1)*B11^2+(C12-1)*C11^2)/B14).

  For the standard error for the difference between two means, $S_W\sqrt{\dfrac{1}{N_1} + \dfrac{1}{N_2}}$:

- Enter in cell B16, the expression =B15*SQRT((1/B12)+(1/C12)).

  Finally, $t_{obt} = \dfrac{(\bar{X}_1 - \bar{X}_2)}{S_W\sqrt{\dfrac{1}{N_1} + \dfrac{1}{N_2}}}$:

- Enter the expression =(B10-C10)/B16 in cell B17.

p-Value for Two Tails

The Excel function T.DIST.2T(tvalue,df) returns directly the two-tail p-value. The function requires as input the ***absolute value*** of t_{obt} and the degrees of freedom; thus,

- In cell B18, we enter =T.DIST.2T(ABS(B17), B14).

Then, for an alpha level of .05, we will reject the null hypothesis of equal means. The "With" mean is significantly larger than the "Without" mean.

Again, Excel has the function T.TEST to produce the *p*-value for one or two tails once we enter the range of the two sets of scores. The function has parameters T.TEST(range1, range2, #tails, type). Where type is 1 for paired groups, and 2 for two groups with equal variance.

- In our case, we type =T.TEST(B2:B7, C2:C9, 2, 2) in cell B26.

Effect Size

The Cohen's d effect size in this case is $d = \dfrac{|\bar{X}_1 - \bar{X}_2|}{S_W}$

- We enter in cell B20, the expression =ABS(D10-C10)/B15. The effect size in this case is large.

Confidence Interval

The 100(1 – α)% CI for the population difference between two independent means is:

$$\left(\bar{X}_1 - \bar{X}_2\right) \pm t_{\alpha/2;N_1+N_2-2}\left(S_W\sqrt{\tfrac{1}{N_1}+\tfrac{1}{N_2}}\right)$$

$t_{\alpha/2;N_1+N_2-2}$ are the critical values from a *t*-distribution with $N_1 + N_2 - 2$ degrees of freedom that defines the upper and lower α/2 areas in the distribution. To obtain the interval half-width $t_{\alpha/2;N_1+N_2-2}\left(S_W\sqrt{\tfrac{1}{N_1}+\tfrac{1}{N_2}}\right)$ we enter the *t* critical values using T.INV.2T (alpha, df) and multiply its value times the standard error that we found before in B16.

- Half-width: Enter in B22 the expression T.INVT.2T(.05,B14)*B16.
- For the lower limit, enter in B23 the expression =(B10-C10)-B22.
- For the upper limit, enter in B24 the expression =(B10-C10)+B22

Thus, possible values for the population difference are between −2.57 and −0.43

Using the Template for Two Independent Groups (Equal Variances)

In the second worksheet in the Excel file for this chapter, we have a template for *t*-test for two independent groups, assuming equal variances (see Figure 14.2). Once we

	A	B	C	D	E	F	G
1		Without	With		t test for Hypothesis about two independent groups		
2		2	4		(Equal variances assumed)		
3		1	3				
4		1	3		H0: μ_1-μ_2 =	0	
5		3	4		Alternative	<>	(<>. <, >)
6		2	3				
7		3	4			Group 1	Group 2
8			2		Sample Mean	2	3.5
9			5		Sample SD	0.89442719	0.9258201
10	Mean	2	3.5		Sample Size	6	8
11	SD	0.894427	0.92582				
12	N	6	8		Degrees of freedom	12	
13					Pooled or Weighted SD	0.91287093	
14							
15					Standard Error	0.49300665	
16					t-obt	-3.04255532	
17					p-value	0.0102252	
18							
19					Effect size	1.64316767	
20							
21					95% C.I.	-2.57416921	-0.4258308
22							

Figure 14.2.

know the averages, standard deviations, and sample sizes for the groups, we can use the template. In the gray cells, we enter the value of the null hypothesis (usually 0), the type of alternative hypothesis ("<>" for two tail, and "<" or ">" for one tail), and the descriptive statistics for the two samples.

Again, using an alpha of .05 we will reject the null hypothesis, since p-value = .0102 (for the two-tail) is smaller than .05. Thus, we can report the results of the test in a paragraph like the following one:

> "Using a t-test for two independent groups, the null hypothesis that the two groups come from populations with the same mean was rejected ($t = -3.043$, df = 12, $p = .0102$). The mean for the "With" group is larger than the mean for the "Without" group. The effect size, $d = 1.643$ is large.

14.3 USING DATA ANALYSIS TOOLPAK FOR *t*-TEST FOR TWO SAMPLES ASSUMING EQUAL VARIANCES

The Data Analysis ToolPak has an option for running a *t*-test for two independent groups when the variances are assumed equal. In the third worksheet in the Excel file for this chapter, we have again the data for the "Without" and "With" groups (see Figure 14.3).

- Click on the Data tab in the top Excel menu, and then click on the Data Analysis icon.
- Select the option "*t*-Test: Two Samples Assuming Equal Variances" and then click OK.
- In the input section, enter as variable 1 the range B1:B7, and as variable 2 the range C1:C9.
- Because we are also selecting the header of the columns, check the options "Labels."
- Enter 0 as the hypothesized mean difference.
- Leave the alpha value at .05.
- Click on "Output Range" and select as first cell E1. Then click OK.

	A	B	C	D	E	F	G
1		Without	With		t-Test: Two-Sample Assuming Equal Variances		
2		2	4				
3		1	3			*Without*	*With*
4		1	3		Mean	2	3.5
5		3	4		Variance	0.8	0.857142857
6		2	3		Observations	6	8
7		3	4		Pooled Variance	0.833333333	
8			2		Hypothesized Mean Difference	0	
9			5		df	12	
10					t Stat	-3.042555317	
11					P(T<=t) one-tail	0.005112602	
12					t Critical one-tail	1.782287556	
13					P(T<=t) two-tail	0.010225204	
14					t Critical two-tail	2.17881283	
15							

Figure 14.3.

WORKED EXAMPLE

14.4

A clinical psychologist thinks that depression may affect sleep. She monitors the sleep of nine depressed patients and eight normal controls for three successive nights. The data for the study is the average number of hours slept by each subject during the last two nights. These values and an empty template are in the four worksheets of the lab file.

Using a *t*-test for two independent groups, check if the two groups *differ* in their average number of sleep hours. Use alpha of .05, and report the conclusion (see Figure 14.4).

1. Compute and report the mean, standard deviation, and sample size for each of the groups (round up the values to two decimals).
   - For means, in cell B11, enter =AVERAGE(B2:B10). Drag the function to cell C11.
   - For standard deviations, in cell B12, enter = STDEV.S(B2:B10). Drag the function to cell C12.
   - For sample sizes, in cell B13, enter =COUNT(B2:B10). Drag the function to cell C13.

For the depressed group, the mean is 6.9 hours, the standard deviation is 0.43 and the sample size is 9. For the not-depressed group, the mean is 7.55 hours, the standard deviation is 0.52 and the sample size is 8.

	A	B	C
1		Deptessed	Not-Depressed
2		7.1	8.2
3		6.8	7.5
4		6.7	7.7
5		7.3	7.8
6		7.5	8
7		6.1	7.4
8		6.9	7.3
9		6.5	6.5
10		7.2	
11	Mean	6.9	7.55
12	SD	0.433012702	0.520988072
13	N	9	8

	E	F	G
1	t test for Hypothesis about two independent groups		
2	(Equal variances assumed)		
3			
4	H0: μ_1-μ_2 =	0	
5	Alternative	<>	(<>. <, >)
6			
7		Group 1	Group 2
8	Sample Mean	6.9	7.55
9	Sample SD	0.4330127	0.520988072
10	Sample Size	9	8
11			
12	Degrees of freedom	15	
13	Pooled or Weighted SD	0.4760952	
14			
15	Standard Error	0.2313407	
16	t-obt	-2.809709	
17	p-value	0.0131986	
18			
19	Effect size	1.3652731	
20			
21	95% C.I.	-1.143091	-0.15690897

t-Test: Two-Sample Assuming Equal Variances		
	Deptessed	*Not-Depressed*
Mean	6.9	7.55
Variance	0.1875	0.271428571
Observations	9	8
Pooled Variance	0.226666667	
Hypothesized Mean Difference	0	
df	15	
t Stat	-2.809708823	
P(T<=t) one-tail	0.006599284	
t Critical one-tail	1.753050356	
P(T<=t) two-tail	0.013198568	
t Critical two-tail	2.131449546	

Figure 14.4.

2. Use the *t*-test for two independent group templates to test the null hypothesis of mean difference 0. For alpha .05, state your conclusion about the null hypothesis.
    - Enter as null hypothesis 0, as type of alternative <>, and the groups' means, standard deviations, and sample sizes.

   For alpha .05, we reject the null hypothesis.

3. Report the 95% CI for the mean difference (round up to two decimals). Would you say that the depressed group sleeps significantly less than the no-depressed group? What is the effect size (again round up to two decimals)? Is this a large or small effect size?

The 95% CI is (−1.14, −.16). Yes, the depressed group sleeps less than the not-depressed group. The effect size is 1.36. This is a large effect size.

4. Repeat the analysis using the Data ToolPak, using also the headers for the Brands.
    - Click on the Data tab in the top Excel menu, and then click on the Data Analysis icon.
    - Select the option "*t*-Test: Two-Sample Assuming Equal Variances" and press OK.
    - In the Input Variable Range 1, enter the range for "Depressed," including the header (B1:B10).
    - In the Input Variable Range 2, enter the range for "Not-depressed," including the header (C1:C10).
    - Type in 0 as the hypothesized mean difference.
    - Check the box "Labels."
    - Leave the alpha value at .05.
    - Select "Output Range" and select as first cell for Table J1. Then click OK.

The final output for the analysis is in worksheet 5 of the chapter Excel file.

15 One-Way ANOVA

Until now we have talked about experimental designs with one or two groups of participants. We can extend experimental designs to more than two experimental conditions or groups. For example, in the previous chapter, a publisher compared the effect of **bold** vs. *italic* highlighting on the recalling of key concepts in a textbook. The publisher can easily add another condition, the ***bold-italic*** condition. In this case, the three types of fonts define the ***levels*** or values of an experimental condition or manipulation. The set of experimental conditions is labeled the experimental ***factor***. Thus, in this case, the design has a single factor "typesetting" with three levels: bold, italic, and bold-italic. Different groups of individuals are assigned at random to each one of the three conditions or levels of the factor. This type of design is called a ***single factor between-subjects design***.

More than two experimental conditions change the way we state the hypotheses. The null hypothesis still states that the sample means come from populations with the same means. However, the alternative hypothesis becomes less specific, because, if we reject the null, we do not know immediately which of all the several group means are different from the others. Thus, a way to represent the hypotheses for a design with K levels or groups is:

$$H_0 : \mu_1 = \mu_2 = \mu_3 = ...\mu_K \qquad \text{vs.} \qquad H_1 : \textit{At least one different}$$

The null hypothesis basically states that all the groups come from the same sampling distribution, i.e., from sampling distributions with the same mean. Of course, even if the group's means come from the same sampling distribution, there will be some variability among them. This variability is quantified by computing the ***variance between*** group's means. This variance compares how similar the group means are to the ***grand mean***, or mean of the group means. The more similar the group means are to each other, the smaller the variance between groups, and the more different the group means are, the larger the variance between groups.

But, there are other consequences of assuming that the groups come from the same sampling distribution. The variability inside each of the groups (i.e., how similar or different are their values) should be the same. These variabilities (or variances) will be the same, and all of them will be estimators of the variance of the common sampling distribution. Thus, we assume that the variances of the groups are the same, and that the combination of these variances, called the ***variance within***, will estimate the variance of the sampling distribution.

Thus, if the null hypothesis of equal group means is correct, we expect the variance between and the variance within to be very similar (or the variance between smaller than the variance within). The test to perform this analysis is called the ***one-way between-subjects analysis of variance***, or simply ***one-way analysis of variance (ANOVA)***. The test-statistic is given by:

$$F_{obt} = \frac{S_B^2}{S_W^2} \quad \text{or} \quad F_{obt} = \frac{MS_B}{MS_W}$$

where the variance between groups' means, S_B^2 or MS_B (notation may differ in different texts) assess how similar or different are the group means, and the variance within groups, S_W^2 or MS_W, assess the variability of the scores within each one of the groups. Each one of these variances has an associated ***degrees of freedom***. The ***degrees of freedom between*** is the number of groups, ***K***, minus 1, i.e., ***K – 1***. The ***degrees of freedom within*** is the total number of participants in the study, N, minus the number of groups *K*, i.e., ***N – K***

When the null hypothesis of equal means is correct, the F_{obt} should have a value of around one (or even less than one). Larger values for F_{obt}, on the other hand, would indicate the rejection of the null hypothesis.

The sampling distribution for the test-statistic F_{obt} is the ***F distribution*** (see Figure 15.1). The *F* distribution is a probability distribution that has as parameters two values of degrees of freedom: The degrees of freedom between or ***degrees of***

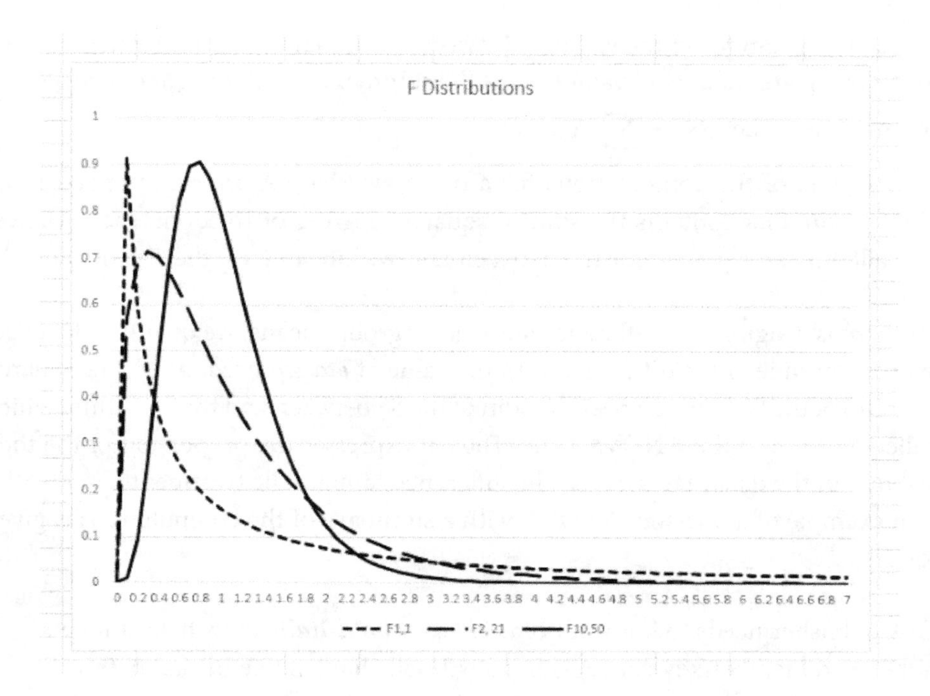

Figure 15.1.

freedom for the numerator in the F ratio, K – 1, is the number of groups minus one. The degrees of freedom within or ***degrees of freedom for the denominator*** in the F ratio, $N - K$, is the total number of observations minus the number of groups.

In comparison to the t-tests, the p-value for ANOVA is defined always as the probability of obtaining the value of F_{obt} or even a larger value, i.e., the p-value is always in the upper tail of F distribution: $p\text{-}value = P(F_{df1, df2} \geq F_{obt})$. In the first worksheet of the Excel for this chapter, we have the graphs of three F distributions: (a) with degrees of freedom 1 and 1, (b) with degrees of freedom 2 and 21, and with degrees of freedom 10 and 50. The distributions are not symmetric, they are usually skewed (i.e., longer right tails) and the F value can only take positive values.

The computation of the test-statistic, F_{obt}, in a one-way ANOVA requires obtaining the variance between and the variance within. However, traditionally we start computing the ***sum of squares*** for the within and between effects. The sum of squares is closely related to the variance, because the variance is the sum of squares divided by the degree of freedoms of the group, i.e., for a variable X, $S_X^2 = SS_X/(N-1)$. Thus, in a one-way ANOVA, we compute first the ***sum of squares between*** and the ***sum of squares within***, and then we divide these Sum of squares by their corresponding degrees of freedom to get the variance between and variance within. Remember that the sum of squares of a set of values, SS, is the summation of their squared deviations from their mean, i.e., $SS_X = \sum_{i=1}^{N}(X_i - \bar{X})^2$.

The results of the computations for a one-way ANOVA are summarized in an ***ANOVA table*** that contains the sum of squares, degrees of freedom, the variances (also called mean squares) for the between and within sources, the F ratio, and the p-value.

If there is a significant difference among the group means, we quantify the effect size or magnitude of the difference with the value of ***eta-squared, or*** η^2. Eta-squared is the ratio of the SS between over the sum of the SS between and the SS within, which is called ***SS total*** in the ANOVA table. Thus, it expresses the proportion of the total variability in the data that is due to the differences among the treatments.

An example of a one-way ANOVA with a summary of the computations is given below:

> A publisher needs to know if **bold**, *italic*, or ***bold-italic*** fonts help students better remember key concepts in a textbook chapter. She prints three versions of the same chapter, one using bold, one using italics, and another using bold-italic to highlight key concepts. She recruits $N = 24$ volunteers and assigns at random an equal number of volunteers to read each version of the chapter ($n = 8$ for each group). Afterwards, she tests the students using a 10-item quiz. For the bold version, the average in the quiz is $\bar{X}_1 = 7.75$ and the standard deviation is $S_1 = 1.25$. For the italic version, the quiz average is $\bar{X}_2 = 6.25$ with standard deviation $S_2 = 1.40$, and, for the bold-italic version the average is $\bar{X}_3 = 8.25$ with standard deviation $S_3 = 1.3$. Are these three means coming from the same sampling distribution, or with sampling distributions with the same mean?

The ***statistical hypotheses*** are: $H_0: \mu_1 = \mu_2 = \mu_3$ vs. H_1: Not all equal

The sum of squares within: $SS_W = \sum_{j=1}^{K}(n_j - 1)S_j^2$

$= (8 - 1)1.25^2 + (8 - 1)1.40^2 + (8-1)1.3^2 = 36.4875$

Degrees of freedom within: $df_W = N - K = (8 + 8 + 8) - 3 = 21$

Variance within: $S_W^2 = SS_W / df_W$ = 36.4875/21 = 1.7375

Grand mean: $\bar{X}_G = \frac{1}{N}\sum_{j=1}^{K} n_j \bar{X}_j$

= (1/24) (8 × 7.75 + 8 × 6.25 + 8 × 8.25) = 7.42

The sum of squares between: $SS_B = \sum_{j=1}^{K} n_j (\bar{X}_j - \bar{X}_G)^2$

= 8(7.75 – 7.42)² + 8(6.25 – 7.42)² + 8(8.25 – 7.42)² = 17.3333

Degrees of freedom between: $df_B = K - 1$ = 3 – 1 = 2

Variance between: $S_B^2 = SS_B / df_B$ = 17.3333/2 = 8.6667

The test-statistic value is: $F_{obt} = S_B^2 / S_W^2$ =8.6667/1.7375 = 4.988

The sampling distribution is: F distribution with 2 df between and 21 df within

The p-value is: Using the *F* distribution

$$p - value = P\left(F_{2,21} \geq 4.988\right) = 0.016886$$

Using ***critical value α*** =.05: Because *p*-value < α, reject the null hypothesis. At least one mean is different.

The ***effect size:*** $\eta^2 = \frac{SS_B}{SS_B + SS_W}$ = 17.3333/(17.3333 + 36.4875) = 0.322056 large effect.

OBJECTIVES

15.1

1. To computer a between-subjects one-way ANOVA using Excel basic functions.
2. To use the chapter template to compute a between-subjects one-way ANOVA.
3. To use the Data ToolPak to compute an "ANOVA: Single Factor" or between-subjects one-way ANOVA.

15.2 ONE-WAY ANOVA USING EXCEL FUNCTIONS

A college professor searches for the best methodology to teach cultural analysis. She decides to try three methods: (1) Lecturing Only, (2) Lecturing plus Activities, and (3) Video demo plus Activities. $N = 27$ volunteers were assigned at random and in equal number, $n = 9$, to one of the three conditions. The same professor taught the three groups and assessed the students using the same mastery test. The professor wants to test the null hypothesis that the three groups come from populations with the same mean, i.e.:

$$H_0 : \mu_1 = \mu_2 = \mu_3 \quad \text{vs.} \quad H_1\text{: Not all the means are the same}$$

The data for the three groups is in the second worksheet of the Excel file for this chapter.

Test Statistic: F_{obt}

First, we compute the means, $\bar{X}_j$, standard deviations, S_j, and sample sizes, n_j, for the $K = 3$ groups (see Figure 15.2).

- In cell B11, we enter =AVERAGE(B2:B10) to obtain the average for the LEC_ONLY condition. Drag the function to cells C11 and D11 to obtain the other two means.
- In cell B12, we enter =STDEV.S(B2:B10) to obtain the standard deviation for LEC_ONLY. Drag the function to cells C12 and D12.
- In cell B13, we enter =COUNT(B2:B10) for the sample size of LEC_ONLY. Drag the function to cells C13 and D13.

Then, we obtain the grand mean, $\bar{X}_G$, i.e., the mean of all observations regardless group.

- In cell F11, enter =AVERAGE(B2:D10). Notice that the range includes all the cells with data.

To obtain the variance within (MSW) we start computing the SS, $(n-1)S^2$, and the degrees of freedom df $= n - 1$, for each group:

	A	B	C	D	E	F
1		LEC_ONLY	LEC_ACT	VID_ACT		
2		91	87	81		
3		85	93	80		
4		86	97	72		
5		76	82	82		
6		80	94	83		
7		87	90	89		
8		91	98	76		
9		83	90	88		
10		84	91	83		
11	Mean	84.77777778	91.33333333	81.55555556	Grand Mean=	85.88888889
12	SD	4.841946349	4.949747468	5.317685378		
13	n	9	9	9		
14	SS	187.5555556	196	226.2222222	SSW=	609.7777778
15	df	8	8	8	dfW=	24
16					MSW=	25.40740741
17						
18	Between	11.11111111	266.7777778	169	SSB=	446.8888889
19					dfB	2
20					MSB	223.4444444
21						
22					Fobt=	8.794460641
23					p=	0.001363964
24						
25					Eta-squared=	0.422923239
26						

Figure 15.2.

- In cell B14, enter the equation = (B13-1)*B12^2, to obtain the sum of square for the LEC_ONLY group. Drag the equation to cells C14 and D14 to get the other groups' SS.
- In cell B15, enter =(B13-1) to get the degrees of freedom for the LEC_ONLY group. Drag the equation to cells C15 and D15.
- In cell F14, enter =SUM(B14:D14), this is the SS within.
- In cell F15, enter =SUM(B15:D15), this is the df within.
- In cell F16, enter the equation =F14/F15, this is the variance or mean square within (MSW).

To obtain the variance between (MSB) we start computing $n(\bar{X} - \bar{X}_G)^2$ for each group:

- In cell B18, enter the formula =B13*(B11-$F$11)^2 to get the value for the LEC_ONLY group (notice that the cell F11, where the grand mean is, has an absolute reference). Drag the equation to cells C18 and D18.
- In cell F18, enter =SUM(B18:D18), this is the SS between.
- In cell F19, enter directly the value of the degrees of freedom Between, $K - 1$, in this example K is 3, thus enter the value 2.
- In cell F20, enter the equation =F18/F19; this is the variance or mean square between (MSB).

The test-statistic is simply the ratio of the variance between over the variance within:

- In cell F22, enter the equation = F20/F16.

p-Value for F_{obt}

The Excel function F.DIST.RT(f.value, df1, df2) returns the right tail probability or area defined by the "f.value" in an F distribution with degrees of freedom "df1" in the numerator and "df2" in the denominator. In other words, the function returns the probability: $P\left(F_{df_1, df_2} \geq F_{obt}\right)$

- In cell F23, enter the function = F.DIST.RT(F22, F19, F15)

In our example, this probability is 0.001364. Thus, for an alpha of .05, we reject the null hypothesis of equal means. At least one of the group means is different from the others.

Effect Size

The effect size in a one-way ANOVA uses "Eta-square": $\eta^2 = SS_B / (SS_B + SS_W)$ This ratio describes the proportion of the total variability in the data that is due to the variability among the treatment conditions. The larger the value, the better.

- In cell F25, enter = F18/(F14+F18).

The eta-squared of 0.423 is considered a large effect size.

When reporting the test-statistic, degrees of freedom and p-value we write: F(2, 24) = 8.79, p = .001364.

ONE-WAY ANOVA USING AN EXCEL TEMPLATE

15.3

We can put the previous one-way ANOVA computations into an Excel template that automatizes the computations. In the third worksheet in the Excel file for this chapter, we have such a template (see Figure 15.3).

	A	B	C	D	E	F	G	H	I
1		LEC_ONLY	LEC_ACT	VID_ACT					
2		91	87	81					
3		85	93	80					
4		86	97	72					
5		76	82	82					
6		80	94	83					
7		87	90	89					
8		91	98	76					
9		83	90	88					
10		84	91	83					
11	Mean	84.77777778	91.33333333	81.55555556					
12	SD	4.841946349	4.949747468	5.317685378					
13	n	9	9	9					
14									
15									
16									
17			1	2	3	4	5	6	7
18		MEANS	84.77777778	91.33333333	81.55556				
19		SD	4.841946349	4.949747468	5.317685				
20		n	9	9	9				
21		SSwj	187.5555556	196	226.2222	0	0	0	0
22		SSbj	11.11111111	266.7777778	169	0	0	0	0
23									
24		K=	3		N=	27			
25		Grand M=	85.88888889						
26									
27			Source	SS	df	MS	F	p	
28			Between	446.8888889	2	223.4444	8.794461	0.001364	
29			Within	609.7777778	24	25.40741			
30			Total	1056.666667	26				
31									
32									
33		Eta-squared	0.422923239						
34									

Figure 15.3.

Again, we compute the means, standard deviations, and sample sizes for the group.

- In cell B11,we enter =AVERAGE(B2:B10) to obtain the average for the LEC_ONLY condition. Drag the function to cells C11 and D11 to obtain the other two means.
- In cell B12, enter =STDEV.S(B2:B10) to obtain the standard deviation for LEC_ONLY. Drag the function to cells C12 and D12.
- In cell B13, enter =COUNT(B2:B10) for the sample size of LEC_ONLY. Drag the function to cells C13 and D13.

We copy the means, standard deviations, and sample sizes to the gray cells in the template:

- Highlight the means, standard deviations, and counts (i.e., the range B11:D13) and copy them to the clipboard (using Ctrl+C or copy).
- Put the cursor on the first gray cell, C18, and right click the mouse. Among the paste options, select the one that copies the **values** of the cells.

15.4 ONE-WAY ANOVA USING DATA ANALYSIS TOOLPAK

The Data Analysis ToolPak provides an option for one-way ANOVA. One of the advantages of this option is that we do not need to compute in advance the mean and standard deviations of the groups.

In worksheet 4 in the Excel file for this chapter, we have again the data and the output from the Data Analysis ToolPak (see Figure 15.4).

- Click the Data tab in the Excel top menu, and then click on the Data Analysis tool icon.
- Select the "Anova: Single Factor" option and then OK.
- As input range, select all the columns with data, including the headers (i.e., the range B1:D10).
- Select "Columns" as the "Grouped by" option.
- Check the box of "Labels in first row."
- Leave the alpha value at .05.
- Click on "Output Range" and select as first cell (upper-left corner of the table) for the output F1, and then click OK.

	A	B	C	D	E	F	G	H	I	J	K	L
1		LEC_ONLY	LEC_ACT	VID_ACT		Anova: Single Factor						
2		91	87	81								
3		85	93	80		SUMMARY						
4		86	97	72		*Groups*	*Count*	*Sum*	*Average*	*Variance*		
5		76	82	82		LEC_ONLY	9	763	84.77778	23.44444		
6		80	94	83		LEC_ACT	9	822	91.33333	24.5		
7		87	90	89		VID_ACT	9	734	81.55556	28.27778		
8		91	98	76								
9		83	90	88								
10		84	91	83		ANOVA						
11						*Source of Variation*	*SS*	*df*	*MS*	*F*	*P-value*	*F crit*
12						Between Groups	446.8889	2	223.4444	8.794461	0.001364	3.402826
13						Within Groups	609.7778	24	25.40741			
14												
15						Total	1056.667	26				
16												

Figure 15.4.

WORKED EXAMPLE

15.5

A researcher conducts a memory experiment to assess if memory changes with age. Four groups that differ in age (30, 40, 50, and 60 years old) but in no other characteristics (such as years of education, IQ, gender, motivation, etc.) participated in the study. Each participant reads three-letter nonsense syllables at a rate of one syllable every 4 seconds. The syllables are shown twice, after which the participants are asked to write down as many of the syllables as they can remember. The data and an empty template are in worksheet 5 in the Excel file for this chapter.

1. Compute and report the mean, standard deviation, and sample size for each of the groups.
   - In cell B8, enter =AVERAGE(B2:B7). Drag the formula to C8, D8, and E8.
   - In cell B9, enter =STDEV.S(B2:B7). Drag the formula to C9, D9, and E9.
   - In cell B10, enter =COUNT(B2:B7). Drag the formula co C10, D10, and E10.

2. Use the one-way ANOVA template to test the null hypothesis of equal means (see Figure 15.5). Report the value of F_{obt}, degree of freedom numerator, degree of freedom denominator, and p-value. For alpha .05, state your conclusion about the null hypothesis.

	A	B	C	D	E	F	G	H
1		30 Year	40 Years	50 years	60 years			
2		14	12	17	13			
3		13	15	14	10			
4		15	16	14	7			
5		17	11	9	8			
6		12	12	13	6			
7		10	18	15	9			
8	Mean	13.5	14	13.66667	8.833333			
9	SD	2.428992	2.75681	2.65832	2.483277			
10	n	6	6	6	6			
11								
12								
13								
14		1	2	3	4	5	6	7
15	MEANS	13.5	14	13.66667	8.833333			
16	SD	2.428992	2.75681	2.65832	2.483277			
17	n	6	6	6	6			
18	SSwj	29.5	38	35.33333	30.83333	0	0	0
19	SSbj	6	13.5	8.166667	80.66667	0	0	0
20								
21	K=	4		N=	24			
22	Grand M=	12.5						
23								
24		Source	SS	df	MS	F	p	
25		Between	108.3333	3	36.11111	5.403159	0.006893	
26		Within	133.6667	20	6.683333			
27		Total	242	23				
28								
29								
30	Eta-squared	0.447658						
31								

Figure 15.5.

- Highlight the means, standard deviations, and counts (i.e., the range B8:E10) and copy them to the clipboard (using Ctrl+C or copy).
- Put the cursor on the first gray cell, B15, and right click the mouse. Among the paste options, select the one that copies the ***values*** of the cells.

$F = 5.4031$, $df1 = 3$, $df2 = 20$, $p = .0069$. Reject the null hypothesis of equal means.

3. Report the effect size (eta-squared) rounded up to four decimals. Is this a large or small effect size?

Eta-squared = 0.4477 This is a large effect size.

4. Obtain the one-way ANOVA using the Data ToolPak (see Figure 15.6).
   - Click the Data tab in the Excel top menu, and then click on the Data Analysis tool icon.

	A	B	C	D	E	F	G
31							
32	Anova: Single Factor						
33							
34	SUMMARY						
35	*Groups*	*Count*	*Sum*	*Average*	*Variance*		
36	30 Year	6	81	13.5	5.9		
37	40 Years	6	84	14	7.6		
38	50 years	6	82	13.66667	7.066667		
39	60 years	6	53	8.833333	6.166667		
40							
41							
42	ANOVA						
43	*Source of Variation*	*SS*	*df*	*MS*	*F*	*P-value*	*F crit*
44	Between Groups	108.3333	3	36.11111	5.403159	0.006893	3.098391
45	Within Groups	133.6667	20	6.683333			
46							
47	Total	242	23				

Figure 15.6.

- Select the "Anova: Simple Factor" option and then OK.
- As input range, select all the columns with data, including the headers (i.e., the range B1:E7).
- Select "Columns" as the "Grouped by" option.
- Check the box of "Labels in first row."
- Leave the alpha value at .05.
- Click on "Output Range" and select as first cell for the output A32, and then click OK.

The result of the analysis is in worksheet 6 in the Excel file for this chapter.

16 One-Way ANOVA: Multiple Comparisons

Rejecting the null hypothesis of equal means in a one-way analysis of variance (ANOVA) only tells us that at least one group mean was different from the others. But, which means are different? To answer this question we need to consider the probability of making a wrong decision in hypothesis testing. We have seen before (Chapter 11) that the critical value alpha is the probability of rejecting the null hypothesis when the null hypothesis is actually true (i.e., the ***Type I error***). When we decide to use an alpha of .05 as the critical value, we are keeping the likelihood of making a Type I error in a *single* test at the level of .05 or less. After rejecting the null hypothesis in a one-way ANOVA, we are bound to run several tests to find which group means are significantly different. Therefore, we need to focus on the risk of making *at least one* Type I error when running several tests. We call the alpha value to test a single test the ***individual alpha error*** and the alpha value to test a set of tests the ***family or experiment-wise alpha error***. In general, keeping the family-wise alpha error at a low value requires the individual alpha errors to be set at even smaller values.

How to test for the different means depends on our research intentions. If before running the one-way ANOVA we have a small number of ***a priori*** hypotheses about the most relevant differences between pairs of means, we can perform a few *t*-tests comparing those pairs of means and use our usual individual alpha errors for the tests. On the other hand, if we do not have any hypotheses about specific differences, we may want to perform ***a posteriori*** comparisons of all possible pairwise differences among the group means. Performing many pairwise comparisons increases dramatically the family-wise alpha error. Fortunately, there are different techniques for performing these multiple comparisons and keep the family-wise alpha error at a desired level. There are several of these techniques that are called in general ***multiple comparison techniques.*** One of the most popular one is the ***Tukey's Honestly Significant Difference (HSD)*** test.

Tukey's HSD ***test-statistic,*** $\boldsymbol{q_{obt}}$, expresses the absolute difference between pair of means in terms of a standard error defined as a function of the variance within from the one-way ANOVA; i.e.,

$$q_{obt} = \frac{\left|\bar{X}_i - \bar{X}_j\right|}{\sqrt{S_W^2/n}}$$

The value of q_{obt} is always positive (the parallel bars in the numerator indicates the absolute value of the difference between the two means). In the denominator of q_{obt} we use the variance between from the one-way ANOVA and the sample size per group, n. If the groups have different sample sizes, the value of n in the q_{obt} equation is replaced by the harmonic mean of the sample sizes.

The sampling distribution for the test-statistic q_{obt} is the ***studentized range distribution.*** This is a probability distribution which is a function of the number of groups to compare, K, and the degrees of freedom within, $df_W = N - K$. Thus, for a desired family-wise alpha, we obtain a critical value for q, q_{crit}, that we use to judge the value of q_{obt}. If the value of q_{obt} is larger than the q_{crit}, the null hypothesis of equal pairs of means is rejected; i.e.,

$$\text{If } q_{obt} > q_{\alpha,K,N-K} \quad \text{reject } H_0$$

Excel does not have a function to obtain values from the studentized range distribution. Thus, we provide a simple table that provides the q critical values for a family-wise alpha of .05 (see Figure16.1).

Using the Tukey's HSD we can complete our ANOVA example in Chapter 15 that compares the recall of students using textbooks with different type of fonts:

	Critical values for q at family-wise alpha .05										
	Number of Group Means (k)										
df_w	2	3	4	5	6	7	8	9	10	11	12
5	3.64	4.60	5.22	5.67	6.03	6.33	6.58	6.80	6.99	7.17	7.32
6	3.46	4.34	4.90	5.30	5.63	5.90	6.12	6.32	6.49	6.65	6.79
7	3.34	4.16	4.68	5.06	5.36	5.61	5.82	6.00	6.16	6.30	6.43
8	3.26	4.04	4.53	4.89	5.17	5.40	5.60	5.77	5.92	6.05	6.18
9	3.20	3.95	4.41	4.76	5.02	5.24	5.43	5.59	5.74	5.87	5.98
10	3.15	3.88	4.33	4.65	4.91	5.12	5.30	5.46	5.60	5.72	5.83
11	3.11	3.82	4.26	4.57	4.82	5.03	5.20	5.35	5.49	5.61	5.71
12	3.08	3.77	4.20	4.51	4.75	4.95	5.12	5.27	5.39	5.51	5.61
13	3.06	3.73	4.15	4.45	4.69	4.88	5.05	5.19	5.32	5.43	5.53
14	3.03	3.70	4.11	4.41	4.64	4.83	4.99	5.13	5.25	5.36	5.46
15	3.01	3.67	4.08	4.37	4.59	4.78	4.94	5.08	5.20	5.31	5.40
16	3.00	3.65	4.05	4.33	4.56	4.74	4.90	5.03	5.15	5.26	5.35
17	2.98	3.63	4.02	4.30	4.52	4.70	4.86	4.99	5.11	5.21	5.31
18	2.97	3.61	4.00	4.28	4.49	4.67	4.82	4.96	5.07	5.17	5.27
19	2.96	3.59	3.98	4.25	4.47	4.65	4.79	4.92	5.04	5.14	5.23
20	2.95	3.58	3.96	4.23	4.45	4.62	4.77	4.90	5.01	5.11	5.20
25	2.91	3.52	3.89	4.15	4.36	4.53	4.67	4.79	4.90	4.99	5.08
30	2.89	3.49	3.85	4.10	4.30	4.46	4.60	4.72	4.82	4.92	5.00
35	2.87	3.46	3.81	4.07	4.26	4.42	4.56	4.67	4.77	4.86	4.95
40	2.86	3.44	3.79	4.04	4.23	4.39	4.52	4.63	4.73	4.82	4.90
45	2.85	3.43	3.77	4.02	4.21	4.36	4.49	4.61	4.70	4.79	4.87
50	2.84	3.42	3.76	4.00	4.19	4.34	4.47	4.58	4.68	4.77	4.85
60	2.83	3.40	3.74	3.98	4.16	4.31	4.44	4.55	4.65	4.73	4.81
100	2.81	3.36	3.70	3.93	4.11	4.26	4.38	4.48	4.58	4.66	4.73
120	2.80	3.36	3.68	3.92	4.10	4.24	4.36	4.47	4.56	4.64	4.71
150	2.79	3.35	3.67	3.90	4.08	4.23	4.35	4.45	4.54	4.62	4.70
Infinity	2.77	3.31	3.63	3.86	4.03	4.17	4.29	4.39	4.47	4.55	4.62

Figure 16.1.

In a study, a publisher compares if **bold**, *italic*, or ***bold-italic*** highlights help students remember key concepts better in a textbook chapter. She prints three versions of the same chapter using each type of highlight and assigns at random $n = 8$ students to each group. Afterward, all students take a 10-item quiz. The means and standard deviations (see Figure 16.2), and the ANOVA table (see Figure 16.3), are given below. The null hypothesis of equal means is rejected. Which group means are different?

	Bold	Italic	Bold & Italic
Mean	7.75	6.25	8.25
SD	1.25	1.4	1.3
n	8	8	8

Figure 16.2.

	SS	df	MS	F	p
Between	17.3333	2	8.6667	4.988	0.016886
Within	36.4875	21	1.7375		
Total	53.8228	23			

Figure 16.3.

Critical value for q: $K = 3$, $df_W = 21$, $\alpha_{Family} = .05$: $q_{crit} = 3.58$ (from the table, approximation using $df_W = 20$).

Bold vs. *italic* $q_{obt} = \frac{|7.75 - 6.25|}{\sqrt{1.7375/8}} = 3.218651$ $q_{obt} < q_{crit}$ Retain hypothesis of equal means.

Bold vs. ***bold-italic*** $q_{obt} = \frac{|7.75 - 8.25|}{\sqrt{1.7375/8}} = 1.072884$ $q_{obt} < q_{crit}$ Retain hypothesis of equal means.

Italic vs. ***bold-italic*** $q_{obt} = \frac{|6.25 - 8.25|}{\sqrt{1.7375/8}} = 4.291534$ $q_{obt} > q_{crit}$ Reject hypothesis of equal means.

Thus, the only significant difference was between the *italic* and the ***bold-italic*** groups. In the Excel file for this chapter, you will find an Excel template to perform these computations.

16.1 OBJECTIVES

1. To compute Tukey's HSD using Excel basic functions.
2. To use the chapter template to compute the between-subjects one-way ANOVA and the Tukey's HSD.

16.2 TUKEY'S HSD USING DATA TOOLPAK AND EXCEL FUNCTIONS

A college professor searches for the best methodology to teach cultural analysis. She decides to try three methods: (1) Lecturing Only, (2) Lecturing plus Activities, and (3) Video plus Activities. $N = 27$ volunteers were assigned at random and in equal number, $n = 9$, to one of the three conditions. The same professor taught the three methods and assessed the students using the same mastery test. The professor wants to test the null hypothesis that the three groups come from sampling distributions with the same mean.

$$H_0: \mu_1 = \mu_2 = \mu_3 \qquad \text{vs.} \qquad H_1\text{: Not all the means are the same}$$

If the null hypothesis is rejected, which groups are significantly different? The data is in the first worksheet of the Excel file for this chapter. To obtain the ANOVA table and the means per groups we will use the Data Analysis ToolPak (see Figure 16.4).

- Click the Data tab in the Excel top menu, and then click on the Data Analysis tool icon.
- Select the "Anova: Single Factor" option and then OK.
- As input range, select all the columns with data, including headers (range B1:D10).
- Select "Columns" as the "Grouped by" option.
- Check the box of "Labels in first row."
- Leave the alpha level at .05.
- Click on "Output Range" and select as first cell for the output F1, and then click OK.

	A	B	C	D	E	F	G	H	I	J	K	L
4		86	97	72		*Groups*	*Count*	*Sum*	*Average*	*Variance*		
5		76	82	82		LEC_ONLY	9	763	84.77777778	23.44444444		
6		80	94	83		LEC_ACT	9	822	91.33333333	24.5		
7		87	90	89		VID_ACT	9	734	81.55555556	28.27777778		
8		91	98	76								
9		83	90	88								
10		84	91	83		ANOVA						
11						*Source of Variation*	*SS*	*df*	*MS*	*F*	*P-value*	*F crit*
12						Between Groups	446.8889	2	223.4444444	8.794460641	0.001363964	3.402826105
13						Within Groups	609.7778	24	25.40740741			
14												
15						Total	1056.667	26				
16												
17												
18	LEC_ONLY vs LEC_ACT			3.90167		Reject equal means						
19												
20	LEC_ONLY vs VID_ACT			1.91777		Retain equal means						
21												
22	LEC_ACT vs VID_ACT			5.819441		Reject equal means						
23												

Figure 16.4.

Thus, the p-value in the ANOVA table is smaller than .05 and we reject the null hypothesis of equal means in the ANOVA. For the Tukey's HSD the q_{crit} for $K = 3$, $df_W = 24$, $\alpha_{Family} = .05$ is approximately $q_{crit} = 3.52$ (using $df_W = 25$ from the table in worksheet 4 of the Excel file for this chapter).

Comparing LEC_ONLY vs. LEC_ACT groups.

- In cell D18, enter the function =ABS(I5-I6)/SQRT(I13/G5). The q_{obt} of 3.90167 is larger than 3.52, we reject the null hypothesis that the two means are the same.

Comparing LEC_ONLY vs. VID_ACT groups.

- In cell D20, enter the function =ABS(I5-I7)/SQRT(I13/G5). The q_{obt} of 1.19177 is smaller than 3.52, we retain the null hypothesis that the two means are the same.

Comparing LEC_ACT vs. VID_ACT groups.

- In cell D22, enter the function =ABS(I6-I7)/SQRT(I13/G5). The q_{obt} of 5.819441 is larger than 3.52, we reject the null hypothesis that the two means are the same.

Thus, we conclude that the LEC_ACT group has a significantly higher mean (91.3) than the other two groups. Also, these two other groups, LEC_ONLY (84.8) and VID_ACT (81.6), have means that are not significantly different.

16.3 EXCEL TEMPLATE FOR ONE-WAY ANOVA AND TUKEY'S HSD

In the third worksheet of the Excel file for this chapter, we have a new template that adds to the one-way ANOVA template a template for the computation of the Tukey's HSD. The template obtains all pairwise differences between means, and the q_{obt} statistics for each pair of means. We compare those values against the q_{crit} from the table in worksheet 4.

The template requires as input the mean, standard deviations, and sample sizes for the groups (see Figure 16.5). These values were obtained in the second worksheet of the Excel file for this chapter.

- In cell B11, we enter =AVERAGE(B2:B10) to obtain the average for the LEC_ONLY condition. Drag the function to cells C11 and D11 to obtain the other two means.

	A	B	C	D
1		LEC_ONLY	LEC_ACT	VID_ACT
2		91	87	81
3		85	93	80
4		86	97	72
5		76	82	82
6		80	94	83
7		87	90	89
8		91	98	76
9		83	90	88
10		84	91	83
11	Mean	84.777778	91.333333	81.55556
12	SD	4.8419463	4.9497475	5.317685
13	n	9	9	9
14				

Figure 16.5.

- In cell B12, enter =STDEV.S(B2:B10) to obtain the standard deviation for LEC_ONLY. Drag the function to cells C12 and D12.
- In cell B13, enter =COUNT(B2:B10) for the sample size of LEC_ONLY. Drag the function to cells C13 and D13.

These values are copied to the third worksheet in the Excel file for this chapter (see Figure 16.6).

- Highlight the means, standard deviations, and counts in the second worksheet (i.e., the range B11:D13) and copy them to the clipboard (using Ctrl+C or copy).
- Go to the third worksheet and put the cursor on the first gray cell for the ANOVA template, C3, and right click the mouse. Among the paste options, select the one that copies the "values" of the cells.

To help you to compare the q_{obt} in the "Studentized q-obtained" table, type in the 0.05 critical q value, 3.52, in cell E32.

Again, Xbar1 and Xbar2, and Xbar2 and Xbar3 are significantly different ($q_{obt} > q_{crit}$).

	A	B	C	D	E	F	G	H	I	J	K	L
1												
2			1	2	3	4	5	6	7	8	9	
3		MEANS	84.77778	91.33333333	81.55556							
4		SD	4.841946	4.949747468	5.317685							
5		n	9	9	9							
6		SSwj	187.5556	196	226.2222	0	0	0	0	0	0	
7		SSbj	11.11111	266.7777778	169	0	0	0	0	0	0	
8												
9		K=	3		N=	27						
10		Grand M=	85.88889									
11												
12			Source	SS	df	MS	F	p				
13			Between	446.8888889	2	223.4444	8.794461	0.001364				
14			Within	609.7777778	24	25.40741						
15			Total	1056.666667	26							
16												
17		Eta-squar	0.422923									
18												
19	Studentized q-obtained											
20			Xbar 2	Xbar 3	Xbar 4	Xbar 5	Xbar 6	Xbar 7	Xbar 8	Xbar 9		
21		Xbar 1	3.902	1.918								
22		Xbar 2		5.819								
23		Xbar 3										
24		Xbar 4										
25		Xbar 5										
26		Xbar 6										
27		Xbar 7										
28		Xbar 8										
29		Xbar 9										
30												
31												
32	For nH=	9	.05 q-critical from table		3.52							
33	For K=	3										
34	dfW=	24										
35	MSw	25.40741										
36	q.den	1.680192										
37												

Figure 16.6.

WORKED EXAMPLE

16.4

To assess whether memory changes with age, a researcher conducts an experiment with four age groups. The groups differ in their age (30, 40, 50, and 60 years old) but in no other characteristics (such as years of education, IQ, gender, motivation, etc.). Each participant is shown three-letter nonsense syllables twice at a rate of one syllable every 4 seconds. After all syllables are presented, the participants are asked to write down as many of the syllables as they can remember.

1. The raw data are in worksheet 5 in the Excel file for this chapter.

   Obtain the mean, standard deviations, and counts for each of the groups (see Figure 16.7).

   - For the averages, enter in B8 the function =AVERAGE(B2:B7). Drag the function to C8, D8, and E8.
   - For the standard deviations, enter in B9 the function =STDEV.S(B2:B7). Drag the function to C9, D9, and E9.
   - For the counts *n*, enter in B10 the function =COUNT(B2:B7). Drag the function to C10, D10, and E10.

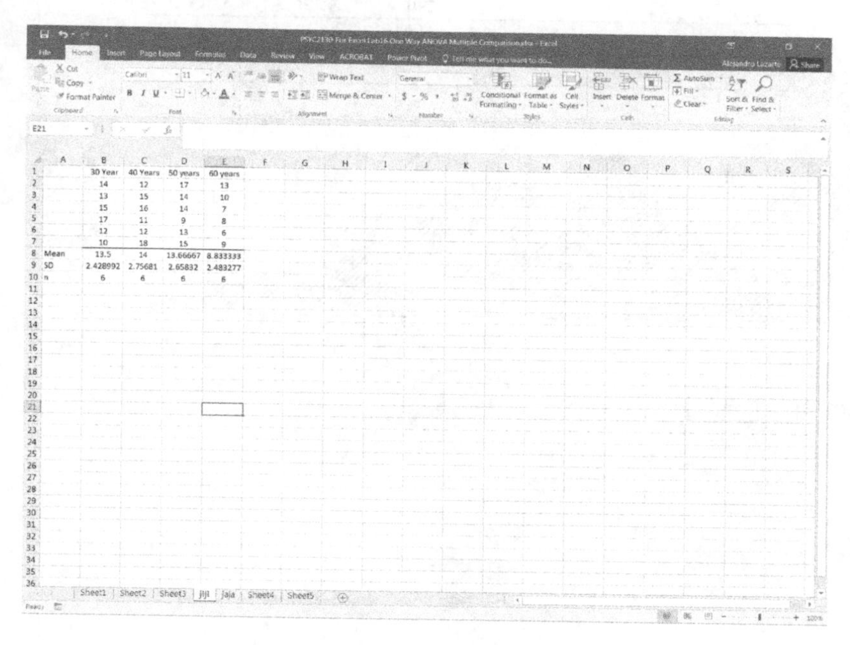

	A	B	C	D	E
1		30 Year	40 Years	50 years	60 years
2		14	12	17	13
3		13	15	14	10
4		15	16	14	7
5		17	11	9	8
6		12	12	13	6
7		10	18	15	9
8	Mean	13.5	14	13.66667	8.833333
9	SD	2.428992	2.75681	2.65832	2.483277
10	n	6	6	6	6

Figure 16.7.

	A	B	C	D	E	F	G	H	I	J	K
1											
2			1	2	3	4	5	6	7	8	9
3		MEANS	13.5	14	13.66667	8.833333					
4		SD	2.428992	2.75681	2.65832	2.483277					
5		n	6	6	6	6					
6		SSwj	29.5	38	35.33333	30.83333	0	0	0	0	0
7		SSbj	6	13.5	8.166667	80.66667	0	0	0	0	0
8											
9		K=	4		N=	24					
10		Grand M=	12.5								
11											
12			Source	SS	df	MS	F	p			
13			Between	108.3333	3	36.11111	5.403159	0.006893			
14			Within	133.6667	20	6.683333					
15			Total	242	23						
16											
17		Eta-square	0.447658								
18											
19	Studentized q-obtained										
20			Xbar 2	Xbar 3	Xbar 4	Xbar 5	Xbar 6	Xbar 7	Xbar 8	Xbar 9	
21		Xbar 1	0.474	0.158	4.422						
22		Xbar 2		0.316	4.895						
23		Xbar 3			4.580						
24		Xbar 4									
25		Xbar 5									
26		Xbar 6									
27		Xbar 7									
28		Xbar 8									
29		Xbar 9									
30											
31											
32	For nH=	6	.05 q-critical from ta		3.96						
33	For K=	4									
34	dfW=	20									
35	MSw	6.683333									
36	q.den	1.055409									
37											

Figure 16.8.

2. Enter the means, SD, and n in the ANOVA and Tukey's HSD template in worksheet 6 in the Excel file for this chapter (see Figure 16.8). Test the null hypothesis of equal means. Report the value of F_{obt}, degree of freedom numerator, degree of freedom denominator, and p-value. For alpha .05, state your conclusion about the null hypothesis.
   - Highlight the means, standard deviations, and counts in worksheet 5 (i.e., the range B8:E10) and copy them to the clipboard (using Ctrl+C or copy).
   - Go to the sixth worksheet and put the cursor on the first gray cell for the ANOVA template, C3, and right click the mouse. Among the paste options, select the one that copies the values of the cells.

   $F = 5.4031$, df1 = 3, df2 = 20, $p = .0069$. Reject the null hypothesis of equal means.

3. Report the effect size (eta-squared) rounded up to four decimals. Is this a large or small effect size?

   Eta-squared = 0.4477. This is a large effect size.

4. Use the Tukey's HSD template output to find which group or groups are significantly different at a family-wise level of 0.05.

For $K = 4$, $df_W = 20$ the value for q_{crit} in table in the fourth worksheet for the present chapter Excel file

The fourth group (the 60-year-olds) is the group that is significantly different from all the other groups.

The results of these analyses are in worksheets 8 and 9 in the Excel file for this chapter.

17 Correlation

We have used tables, graphs, and numerical indicators to describe data from single variables. For example, we learned how to describe the distribution of American College Testing (ACT) scores, to find the average of the number of hours watching cable TV, or to compute the standard deviation for college freshman grade point average (GPA). In addition, we used inferential statistical procedures to test for the equality of one or more sample means and to find confidence intervals (CIs) for the means. Now, we will start addressing a very important question in data analysis: How to describe the relationship between two variables. For example, is there any relationship between the number of hours watching cable TV and the GPA of college students? Questions about the ***relationship between variables*** are pervasive in science, for example:

- Is there any relation between the height of a person and his/her weight?
- Does the number of hours slept the night before help or hinder the performance in a math test?
- Is it true that the faster we read, the worse our recall will be of the material we read?
- Does an extrovert personality influence the job performance of bank tellers?
- Can we predict students' GPA at the end of the first year of college by using their ACT scores?

To answer questions about relations we have two, closely related, procedures that focus on detecting ***linear*** relations between two variables: correlation and regression.

> A ***correlation coefficient*** is a *single value* that expresses the degree and strength of a linear relationship between two variables. It is a single value that can range between −1, for a perfect linear negative relation, to 1 for a perfect positive relation. A correlation coefficient close to zero would indicate a lack of linear relation between the two variables.
>
> A ***linear regression*** is an *equation* that describes the relation between a predictor variable X and a response variable Y by using as a model the equation of a line.

In this chapter, we introduce the basic ideas about correlation and how we can use Excel to obtain the required computations.

Correlation is intended to detect *linear* relations. Thus, the first step on correlational analysis is to have a general idea of this linearity. The basic plot for describing the relation between two variables is the ***scatterplot***. For example, suppose that X is the number of hours slept the night before a big math exam and Y is the score in the exam. On a two-axis plot, where X represents the number of sleep hours and Y the math exam scores, we mark the pairs (X, Y) for each of participants. There are two scatterplots below: One of them shows data with a potential *linear* relation between X and Y (see Figure 17.1). The cloud of points seems to have a positive trend, i.e., the larger the number of hours slept, the higher the score in the math test. On the other

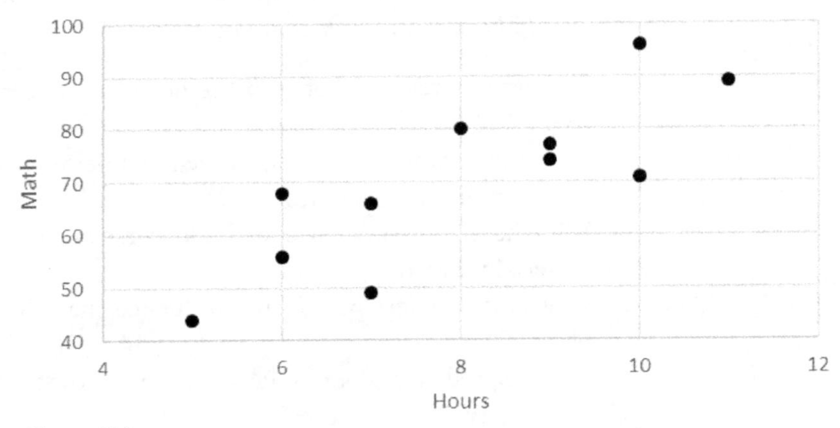

Figure 17.1.

hand, the other data plot shows a potential *nonlinear* relation between X and Y (see Figure 17.2). The performance on the math test increases with the number of sleeping hours ... but only to a certain point. After eight hours of sleep, the math performance tends to decrease. Correlation will be high for the linear data, but not so high with the nonlinear data, even when the nonlinear pattern seems quite good.

The data showing a linear trend for sleep and math suggests that both have good linear relation. But we need to quantify this perception. The first value to perform this quantification is the ***covariance*** between X and Y. Covariance takes positive and negative values. If the covariance is large and negative, we have a negative linear trend. If the covariance is large and positive, we have a positive trend. If the covariance is around zero, we do not have any linear trend. The equation for the covariance between X and Y is represented by $\boldsymbol{S_{XY}}$ and its formula is:

$$S_{XY} = \frac{\sum_{i=1}^{N}(X_i - \bar{X})(Y_i - \bar{Y})}{N-1}$$

where the numerator has its own name. It is called the ***sum of cross products*** of X and Y and is represented by the expression $\boldsymbol{SS_{XY}}$.

Although the value of the covariance measures relationship between two variables, the actual value of the covariance is unbounded, i.e., we don't know what is a

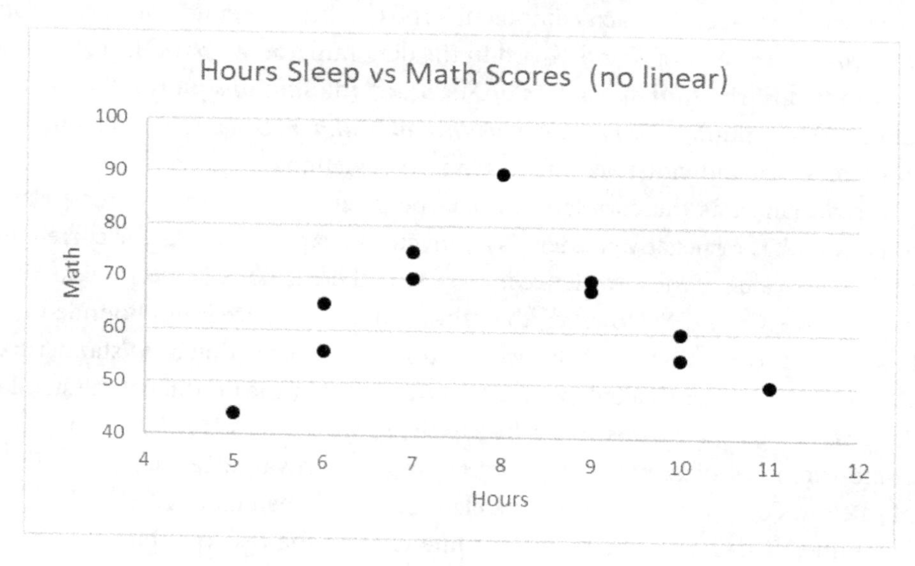

Figure 17.2.

large or a small covariance. We need an index that is bounded, i.e., we know what its maximum and minimum values are. The most popular index for this purpose is the ***coefficient of correlation r***, or more precisely, the *Pearson Product-Moment Correlation Coefficient r*. The correlation coefficient ranges between −1 and 1. Thus, the sign of r tells us the directionality of the relationship. When r is negative, X and Y are inversely related (when X increases, Y decreases), while when r is positive, X and Y are directly related (when X increases, Y increases). The absolute value of r, tells us the strength of the relationship between the two variables. The closer the absolute value of r is to 1, the stronger is the relation, and the closer the absolute value of r is to 0, the weaker the relation is.

There are alternative ways to write the formula for the coefficient correlation between two variables X and Y. Three of them are depicted below:

$$(1)\ r = \frac{\sum_{i=1}^{N}(X_i - \bar{X})(Y_i - \bar{Y})}{\sqrt{\sum_{i=1}^{N}(X_i - \bar{X})^2 \sum_{i=1}^{N}(Y_i - \bar{Y})^2}} \qquad (2)\ r = \frac{SS_{XY}}{\sqrt{SS_X\, SS_Y}} \qquad (3)\ r = \frac{S_{XY}}{S_X S_Y}$$

Equation (1) displays the formula as the sum of the cross products for X and Y divided by the square root of the product of the sum of squares of X times the sum of squares of Y. Equation (2) is the same formula, but we replace the summation notation with the labels for each component. Thus, in the numerator we have the ***sum of cross-products***, SS_{XY}, of X and Y, and in the denominator, we have the square root of the product of the ***sum of squares of X***, SS_X, by the ***sum of squares***, SS_Y, **of *Y***. In equation (3), the numerator is the ***covariance of X and Y***, S_{XY}, and the denominator the product of the individual variables standard deviations.

The covariance, as the correlation r, can be positive if there is a direct relation between X and Y, or negative if there is an inverse linear relation. As the correlation, if the absolute value of the covariance is close to 0, there is none or a very weak linear relation between X and Y. However, the value of the covariance is not bounded to −1 or 1 as it is the correlation. Actually, we can think of a correlation as a "standardized covariance" in the sense that by dividing the covariance by the product of the standard deviations, we restrict the possible values to the range of −1 to 1.

Some examples of scatterplots, correlations, and covariances are given in the graphs below. For a perfect positive correlation, $r = +1$ (see Figure 17.3) , all the dots fall on a line. This line is the regression line relating the two variables. For a not perfectly positive correlation (e.g. $r = 0.83$ in Figure 17-4), the dots are scattered

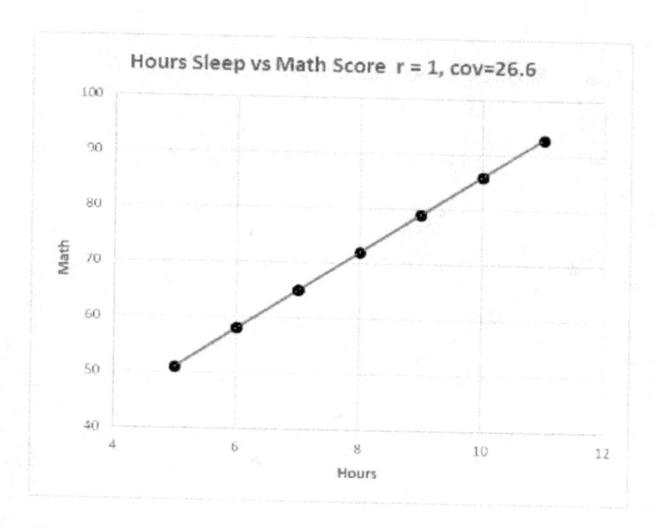

Figure 17.3.

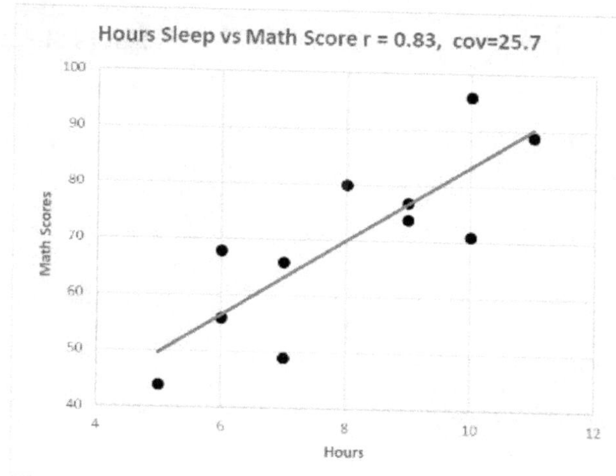

Figure 17.4.

around the regression line. When there is no correlation between the variables, i.e. $r = 0$ (see Figure 17-5) the slope of the regression line is zero and the line is horizontal. When we have a not perfectly negative correlation (e.g. r = -0.74 in Figure 17-6) the dots are scattered around a line with a negative trend. Finally, for a perfect negative correlation ($r = -1$, see Figure 17.7) all the dots fall on a line with negative slope. We will talk more about regression in the next chapter.

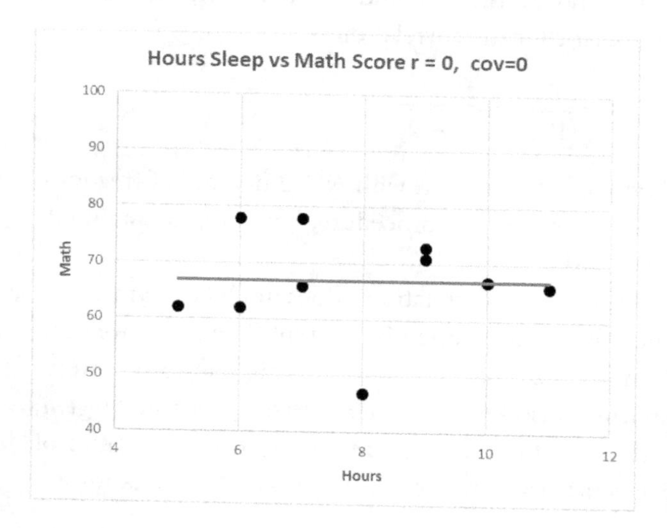

Figure 17.5.

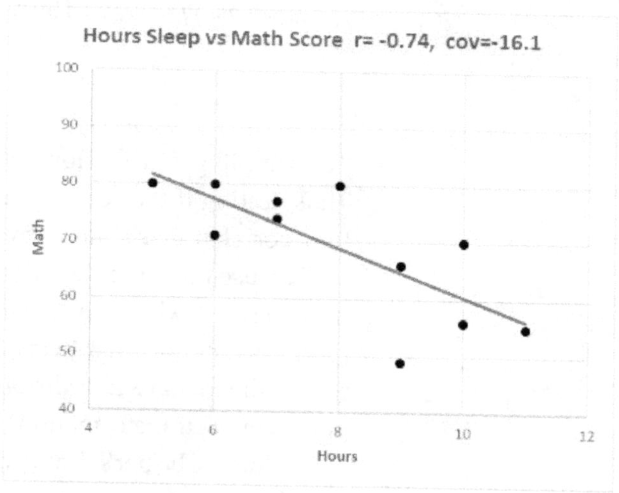

Figure 17.6.

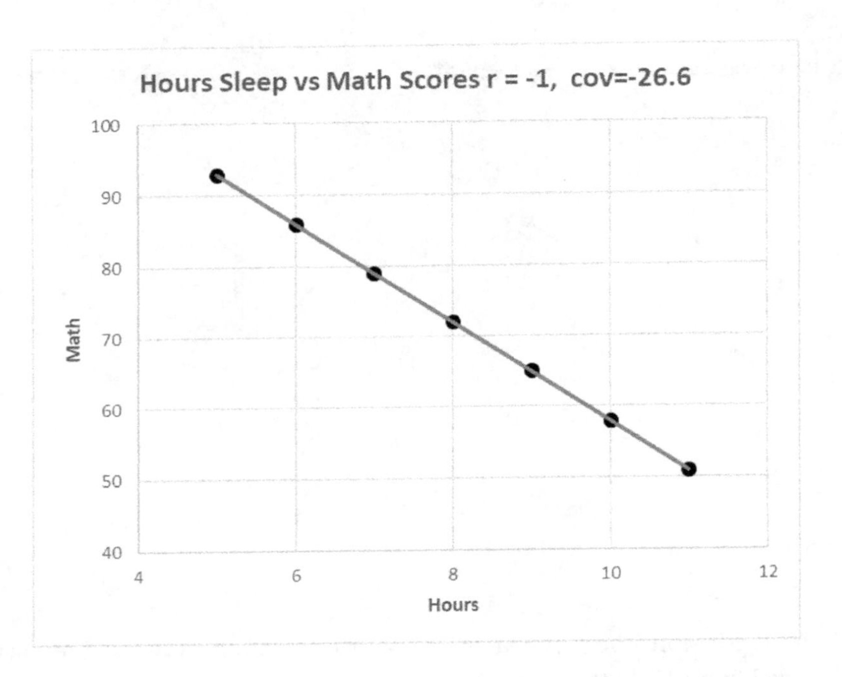

Figure 17.7.

In terms of inferential procedures, we test if a sample correlation is coming from a population where the actual correlation is zero. If we reject this null hypothesis, we say that the sample correlation is significantly different from zero and therefore indicates a relationship between X and Y. The null and alternative hypothesis are: $H_0; \rho = 0$ vs. $H_1 : \rho \neq 0$. The test-statistic for the correlation is:

$$t_{obt} = \frac{r}{\sqrt{(1-r^2)/(N-2)}}$$

The sampling distribution will be the t distribution with N – 2 degrees of freedom. Thus, testing if the correlation is zero is a simple procedure; however, to obtain CIs for a correlation is a more involved task.

Because the sampling distribution for the correlation r does not take a symmetric normal shape when the actual population correlation is different than $\rho = 0$, we need to use a function that helps to normalize that distribution. This one-to-one function is the ***Fisher's Z transformation*** (also known as the *inverse hyperbolic tangent*, or *arctanh*, transformation). This function, and its inverse, transforms values of r to values of Fisher's Z and vice versa. To compute the 95% CI for a correlation we actually follow three steps:

First, we transform the sample value of r to its Fisher's Z transformed value, $Fz(r)$. Second, we obtain the upper and lower limits of the CI by using:

$$Lower\ Fz(r):\ Fz(r) - 1.96\sqrt{1/(N-3)} \qquad Upper\ Fz(r):\ Fz(r) + 1.96\sqrt{1/(N-3)}$$

Third, we transform back the upper and lower limits to correlation values using the inverse Fisher's Z transformation.

OBJECTIVES

17.1

1. To compute covariance and correlation using Excel functions.
2. To compute correlation, test of significance, and confidence for a single correlation using an Excel template.
3. To compute several correlations at once using the Data Analysis ToolPak.
4. To compute test of significance for several correlations at once using an Excel template.

SINGLE CORRELATION, TEST, AND CONFIDENCE INTERVAL USING EXCEL BASIC FUNCTIONS AND AN EXCEL TEMPLATE

17.2

In the first worksheet of the Excel file for this chapter we have information about the height of nine students, X, and the height of their mothers (Y) (see Figure 17.8). We first obtain the mean, standard deviation, and sample size for both variables.

- In cell B11, enter =AVERAGE(B2:B10). Drag the formula to cell C11 too.
- In cell B12, enter = STEDV.S(B2:B10). Drag the formula to cell C12 too.
- In cell B13, enter = COUNT(B2:B10). Drag the formula to cell C13.

Excel has functions to obtain the covariance and correlation of two range of values: COVARIANCE(range1, range2), and CORREL(range1, range2). In our example:

- In cell B15, enter = COVARIANCE(B2:B10, C2:C10)
- In cell B16, enter = CORREL(B2:B10, C2:C10)

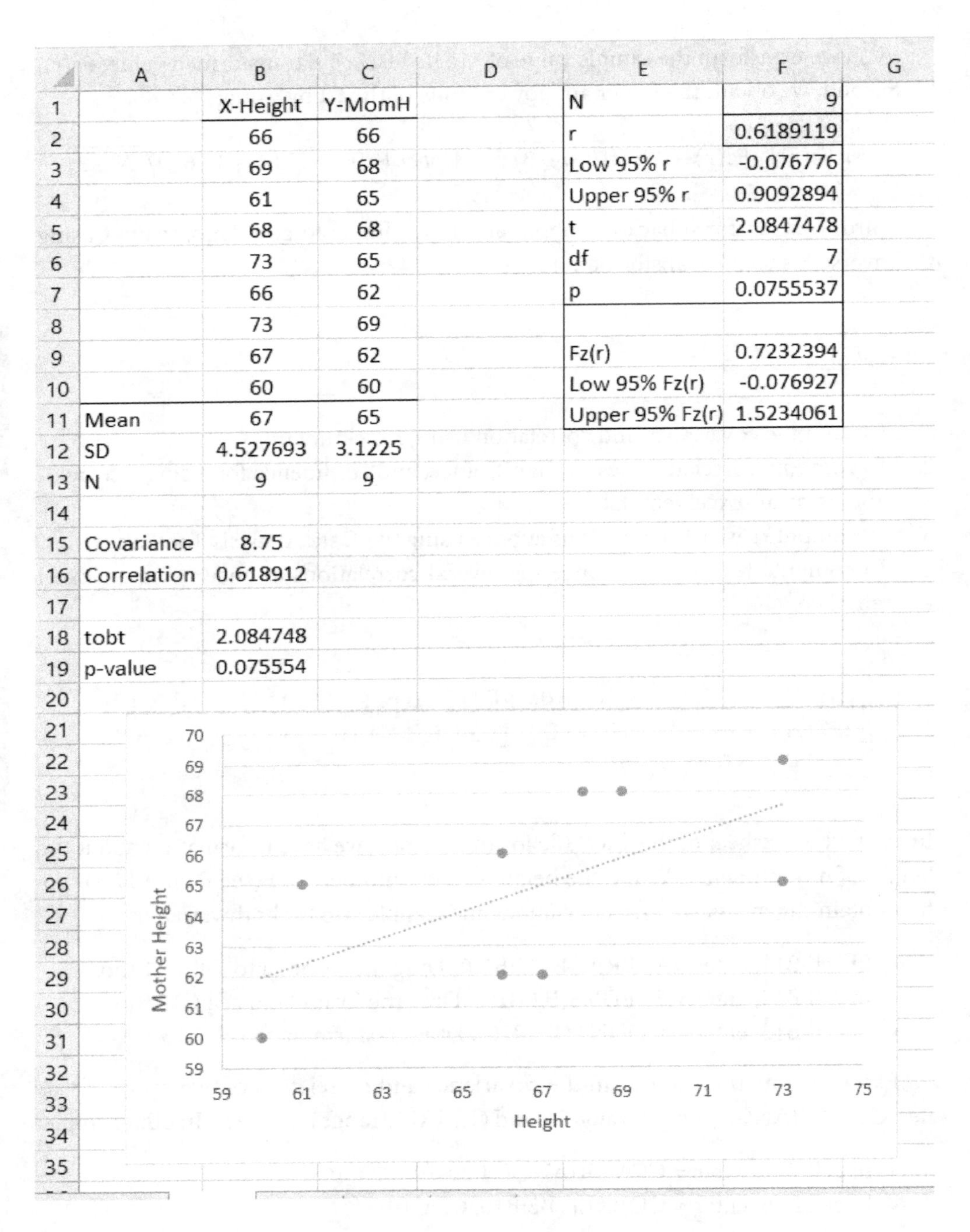

	A	B	C	D	E	F	G
1		X-Height	Y-MomH		N	9	
2		66	66		r	0.6189119	
3		69	68		Low 95% r	-0.076776	
4		61	65		Upper 95% r	0.9092894	
5		68	68		t	2.0847478	
6		73	65		df	7	
7		66	62		p	0.0755537	
8		73	69				
9		67	62		Fz(r)	0.7232394	
10		60	60		Low 95% Fz(r)	-0.076927	
11	Mean	67	65		Upper 95% Fz(r)	1.5234061	
12	SD	4.527693	3.1225				
13	N	9	9				
14							
15	Covariance	8.75					
16	Correlation	0.618912					
17							
18	tobt	2.084748					
19	p-value	0.075554					
20							
21							
22							
23							
24							
25							
26							
27							
28							
29							
30							
31							
32							
33							
34							
35							

Figure 17.8.

The correlation of 0.6189118 is positive and quite high. Excel does not have a basic function to test this correlation; however, we can compute the test using the formulas given above. First, we compute the value of t_{obt}. Remember that the value of the correlation, r, is in B16, and the value of the sample size, N, can be obtained from B13. Thus:

- Enter in cell B18 =B16/SQRT((1-B16^2)/(B13-2))

The t_{obt} value is 2.044748. We need to obtain the p-value for this t_{obt} with degrees of freedom $N - 2$. We use the function T.DIST.2T

- Enter in cell B19 =T.DIST.2T(B18, B13-2)

The p-value is .075554. Thus, for an alpha of .05 we cannot reject the null hypothesis that the sample correlation comes from a population with correlation zero.

We can obtain the corresponding scatterplot with the regression line trend.

- Highlight columns B and C containing the data, including the headers (range B1:C10).
- Click the Insert tab in the Excel top menu. In the Charts click on the Scatter icon and select the Scatter option. This will create the basic scatterplot.
- In the Excel top menu, you will see a Chart Tools tab with two options: Design and Format. Click on the Design tab and the "Quick Layout" icon. Select the first layout option. This will allow you to enter titles for the x (height) and y (mothers' height) axes. Erase the overall title of the chart (the Y-Absent label) and the legend on the right-hand side of the plot.
- Excel tends to display axis starting at zero, such as the one for the horizontal axis Height. To redefine the axis, click twice on the horizontal axis values. This will open the "Format Axis" menu.
- In the Format Axis menu, click on the bars icon and click on the Axis Options. Enter as minimum value 59, to overlay the regression line.
- Click on the scatterplot and on the Plus icon that will appear on the right-hand side of the plot.
- Check the option "Trendline" (the regression line will appear on the scatterplots).

In the same first worksheet, we have a simple template for testing if the sample correlation comes from a population with correlation zero. The template, in addition, also computes the 95% CI for the population correlation.

- Enter in the gray cells of the template the sample size, N, and the correlation, r. We can copy the value of the correlation in cell B16 and paste it (as a value, not as a function) in the template.

The template reports a t_{obt} of 2.0847478, with 7 degrees of freedom, and a p-value of .0755537. Therefore, at an alpha of .05 we cannot reject the null hypothesis that the sample correlation comes from a population with correlation zero. In other words, the correlation is not statistically significant. Although a correlation of 0.6189119 seems a sizable one, the small sample size (only 9 participants) makes the p-value not small enough.

The template reports the 95% CI for the correlation (reported in the cells ***Low 95% r*** and ***Upper 95% r***). In agreement with our decision of not rejecting the null hypothesis, the 95% CI for the population correlation includes as possible population correlation the value of 0.

In addition, the template reports the Fisher's Z transformation of the correlation, Fz(r), and the low and upper intervals for the Fisher's Z.

17.3 MATRIX OF CORRELATIONS USING TOOLPAK AND TEMPLATE FOR TESTING CORRELATIONS

Most of the times, researchers want to study the relations between more than two variables. For example, in the second worksheet of the Excel file for this chapter, we have data for 96 undergraduate students' height and weight, as well as their mother's and father's heights. We can study all possible relations among these four variables. One of the advantages of the correlation coefficient is that we can compute and display many correlations in a concise manner using a ***matrix of correlations*** (see Figure 17.9). A matrix of correlations for K variables is a $K \times K$ table that displays the correlations between all pairs of variables. The diagonal entries in the table are always 1, because the correlation of a variable with itself is always a perfect 1. The entries above and below the diagonal are the same, because the correlation between X and Y is the same value as the correlation between Y and X. Thus, usually we only report the upper or lower parts of the correlation matrices.

To obtain the correlation matrix for all pairs of variables, we use the Data Analysis ToolPak.

- Click on the Data tab in the Excel top menu and then on the Data Analysis icon.
- Select the option "Correlation" and then click OK.

	A	B	C	D
1	Height	mom_height	dad_height	weight
2	66	66	70	121
3	69	68	73	140
4	61	65	68	125
5	68	68	70	135
6	73	65	70	170
7	66	62	70	130
8	73	65	70	170
9	67	62	72	125
10	60	60	72	120
11	62	67	67	102
12	65	64	72	135
13	75	61	73	215
14	64	66	72	120
15	68	60	70	210
16	64	66	72	120
17	70	64	64	140
18	74	64	73	220
19	74	65	68	150
20	65	65	68	140
21	71	66	71	185
22	54	56	59	115
23	72	68	72	155
24	69	65	72	130
25	76	69	68	220
26	61	65	69	110
27	66	65	73	145
28	76	69	68	220
29	60	62	65	120
30	68	64	80	132
31	67	72	76	115
32	67	72	76	115
33	66	60	69	220

	Height	*mom_height*	*dad_height*	*weight*
Height	1			
mom_height	0.360940789	1		
dad_height	0.356010602	0.41412689	1	
weight	0.678756444	-0.01587513	-0.0111145	1

r

	1	2	3	4
1	1			
2	0.360940789	1		
3	0.356010602	0.41412689	1	
4	0.678756444	-0.01587513	-0.0111145	1
N=	96			

tobt

	1	2	3	4
1	1			
2	3.752405172	1		
3	3.693651844	4.41114553	1	
4	8.961213686	-0.15393447	-0.1077655	1
df=	94			

p

	1	2	3	4
1	1			
2	0.000302975	1		
3	0.000371048	2.7362E-05	1	
4	2.97167E-14	0.87799141	0.91441137	1

Figure 17.9.

- Put as input range all the data in columns A to D (include headers), i.e., the range A1:D97.
- Check the "Labels in First Row" box.
- Check "Output Range" and select as first cell F1, then press OK.

Notice that the cells on the diagonal of correlation table contain 1.00. In addition, the cells above the diagonal are empty because they are repetitious. For example, if we read the matrix by rows, the correlation between mom_height and Height (0.360940789) will be the same as the correlation between Height and mom_height, i.e., left empty.

Students' height and weight show the largest positive correlation (0.68 if we round it to two decimals), suggesting a positive linear trend between these two variables. On the other hand, students' weight and dad_height show the smallest correlation (regardless sign), i.e., a correlation close to zero (−0.02), suggesting no linear relationship between these two variables.

In the same worksheet, we have a template to find the t_{obt} and p values for testing the sample correlations in the 4 × 4 matrix.

- Enter the correlations in the gray cells in the first table of the template (the one with r as header). You can copy and paste the correlations generated by the Data Analysis ToolPak.
- Enter the sample size N in the corresponding gray cell. The template will compute and display in the second table, the test-statistic t_{obt} and the degrees of freedom for the test. In the third table, the template displays the p values for each test.

Thus, for alpha .05, we have only two correlations that are not significantly different from zero: the correlation between students' weight and mothers' height (p = .87799141) and the correlation between students' weight and father height (p = .91441137).

17.4 WORKED EXAMPLE

Usually research hypotheses take the form of statements about correlations. For example, in one of his manuscripts about the human body, Leonardo da Vinci wrote,"... the span to which the man opens his arms is equivalent to his height." Using our knowledge of correlation, we can translate this statement into, "The span to which individuals open their arms is positively correlated to their heights." If this is correct, we should find a large, and significant, positive correlation between armspan and height. Of course, other body measures may also correlate with height and armspan.

	A	B	C	D	E	F	G	H	I
1	head_in	height_in	arm-span					N	65
2	23.23	66	66		Height & Arm Span			r	0.968684404
3	21.26	64	64		r=	0.968684		Low 95% r	0.948995924
4	22.24	64	63		N=	65		Upper 95% r	0.980847467
5	21.06	67	67					t	30.96593468
6	22.44	65	64					df	63
7	21.85	63	64					p	7.88742E-40
8	21.85	70	69						
9	22.44	71	71					Fz(r)	2.070502273
10	21.26	62	61					Low 95% Fz(r)	1.821582024
11	20.87	67	67					Upper 95% Fz(r)	2.319422522
12	22.05	64	65						
13	22.05	70	68						

Figure 17.10.

In the third worksheet in the Excel file for this chapter, we have three body measures—the head circumference, the height, and the armspan—for 65 students. The worksheet also contains the templates for testing a single correlation and for testing a matrix of correlations (see Figure 17.10).

1. Find the correlation between height and armspan and use the template for a single correlation to test if the correlation is significantly different from zero. (a) Report the value of the correlation, the t_{obt}, degrees of freedom, and the p-value (round up to two decimals the correlation, and up to four decimals the t and p values). Is the correlation significant? (b) Report the 95% CI for the correlation (round up to two decimals). (c) What happens to the armspan when the height of the person increases? Does the data support Leonardo's statement?
   - In cell F3, we find the correlation by entering =CORREL(B2:B66, C2:C66).
   - In cell F4, we find the number of observations by entering =COUNT(C2:C66).
   - We copy the value of N and the correlation to the gray cells in the template.

   a) The correlation is $r = 0.97$, $t = 30.9659$, and $p < .0001$. The correlation is significant.

   b) The 95% CI for the correlation is (0.95, 0.98).

   c) Because the correlation is positive and quite large (i.e., very close to 1), when the height of the person increases, the armspan also increases. Yes, the data supports Leonardo's statement.

	A	B	C	D	E	F	G	H	I	J
13	22.05	70	68							
14	21.65	70	70							
15	21.65	65	65			*head_in*	*height_in*	*arm-span*		
16	22.05	68	68		head_in	1				
17	25.2	66	67		height_in	0.463975	1			
18	21.65	68	68		arm-span	0.50055	0.968684	1		
19	22.05	64	63							
20	21.65	69	69							
21	21.65	64	62							
22	20.87	66	67		r					
23	22.44	62	63			1	2	3	4	
24	24.02	74	75		1	1				
25	23.43	69	70		2	0.463975	1			
26	23.03	65	66		3	0.50055	0.968684	1		
27	23.23	69	70		4				1	
28	22.05	67	67		N=	65				
29	23.62	63	63							
30	22.05	66	67		tobt					
31	22.05	70	70			1	2	3	4	
32	22.44	67	67		1	1				
33	21.65	67	66		2	4.157243	1			
34	22.05	73	74		3	4.589299	30.96593	1		
35	22.05	65	64		4	0	0	0	1	
36	20.47	64	64		df=	63				
37	23.82	67	66							
38	22.44	65	66		p					
39	21.26	62	63			1	2	3	4	
40	21.65	64	63		1	1				
41	22.44	67	68		2	9.91E-05	1			
42	22.83	64	64		3	2.17E-05	7.89E-40	1		
43	23.23	71	71		4	1	1	1	1	
44	22.05	63	63							
45	22.05	64	64							

Figure 17.11.

2. Use the Data Analysis ToolPak, to find the correlation among the three variables. Also, use the template for obtaining the p values for the correlation matrix (see Figure 17.11). (a) Which pair of variables has correlations significantly different from zero (use alpha of .05)? (b)Which variable correlates better with armspan—height or head size? Report the value of the correlation and the p-value (round up to two decimals the correlation and up to four decimals for the p-value).
   - Click on the Data tab in the Excel top menu and then on the Data Analysis icon.
   - Select the option "Correlation" and then click OK.
   - Put as input range all the data in columns A to C (include headers), i.e., the range A1:C66.
   - Check the "Labels in First Row" box.
   - Check "Output Range," select as first cell (i.e., the upper-left corner of the table) E15, and then press OK.
   - Copy the correlations to the gray cells in the template.
   - Enter the value of N in the template.

   a) All the correlations are significantly different from zero.
   b) Height correlates better with armspan: $r = 0.97$, $p < .0001$.

The results of the analysis is in worksheet 4 of the Excel file for this chapter.

18 Simple Linear Regression

In comparison with correlational analysis, linear regression analysis provides us with a *regression equation* to predict values for a response variable Y given the value of a predictor variable X. For example, what will be the estimated weight (Y) of an 18- to 26-year-old male when we know his height (X)? Of course, to use the regression equation we need to test if there is a statistically significant relation between X and Y (by testing the regression slope) and to assess how important the predictor X is in affecting the response Y (by computing the r-squared). The questions addressed with regression analysis are quite similar to the ones for correlational analysis. For example, we would like to answer questions such as the ones below:

- Is there any relation between students American College Testing (ACT) scores and their freshman grade point average (GPA)?
- Are extroverted people better salespeople?
- Is it true that wealth and happiness go together?
- Does individuals' confidence in their memory relate to their actual recall of events?
- Does intelligence relate to professional success?
- Does the amount of iron in children's diet affect their subsequent cognitive development?

To answer questions about relations, we use mathematical functions that describe the relation between a ***predictor variable***, X and a ***response***

variable, Y. Several mathematical functions may help us to model the relation between these two variables X and Y, but the simplest of them all is the ***equation of a line***. You probably have seen this equation before in the following form:

$$Y = mX + b$$

In this equation, ***slope m*** tells us how much the value of the response Y changes when the value X changes by one unit. For example, when slope m is 2, the increase of one point in X produces an *increase* of 2 points in Y. However, if slope m is -4, then an increase of one point in X produces a *reduction* of 4 points in Y, because the slope is negative. ***Intercept b*** in the equation is the value of the response Y when the value of the predictor X is zero. Once we have the slope and intercept of the equation, we can plot a perfect line, i.e., one in which all pairs of values (X, Y) are always on the line. For example, suppose that X is the number of hours slept the night before a big math exam and Y is the score in the exam. Obtaining a line relating those two variables will be quite easy if the data and scatterplot look as shown in Figure 18.1.

In other words, the two variables have a perfect linear relation. All the dots, representing the points (X, Y) in the data, are exactly on the same line. However, in real life data is not so well-behaved. Actually, a more realistic data for the number of hours slept and math scores is displayed in Figure 18.2.

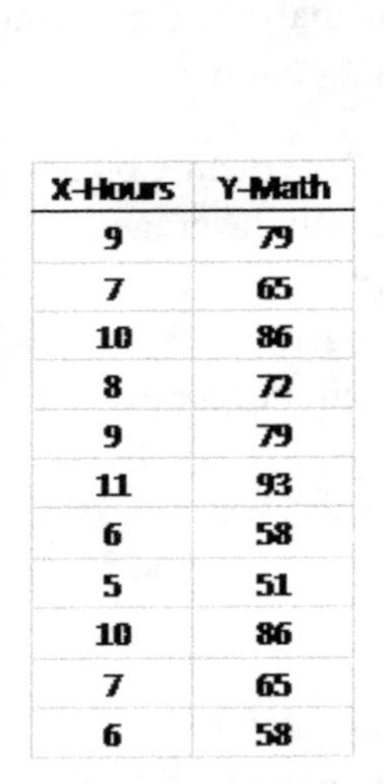

X-Hours	Y-Math
9	79
7	65
10	86
8	72
9	79
11	93
6	58
5	51
10	86
7	65
6	58

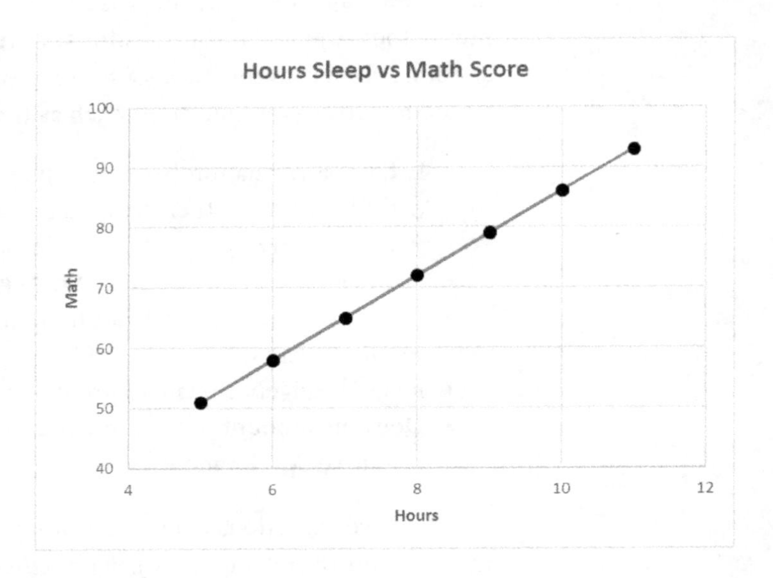

Figure 18.1.

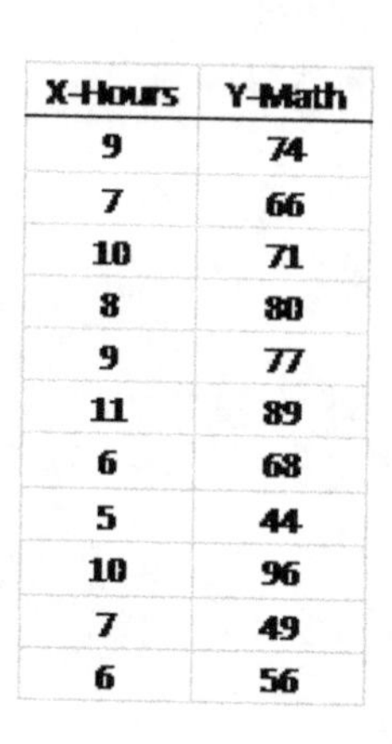

X-Hours	Y-Math
9	74
7	66
10	71
8	80
9	77
11	89
6	68
5	44
10	96
7	49
6	56

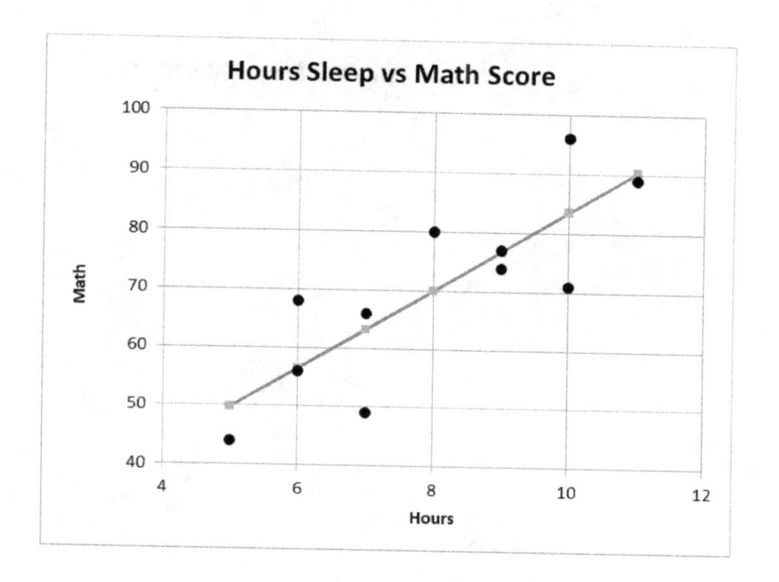

Figure 18.2.

The individual points (*X*, *Y*) do not lie on a perfect line. The ***scatterplot*** suggests a positive trend relating hours of sleep and math scores, because the longer the hours of sleep, the higher the math score. However, not all the (*X*, *Y*) points can be located on a single line. The best we can do is to find a line that is "close enough" to most of the points in the scatterplot. The vertical departures of the points from the line are called ***errors of prediction*** or ***residuals*** (see Figure 18.3). Some of these residuals are positive, i.e., the points are above the line, and others are negative, i.e., below the line. Thus, a "close enough" line is one that minimizes the sum of squared residuals, or ***least square regression line.***

If we know all the values in a population, the equation representing the *population **simple regression line equation*** between *X* and *Y* is the following:

$$Y_i = \beta_0 + \beta_1 X_i + \varepsilon_i \qquad VAR(\varepsilon_i) = \sigma_\varepsilon^2$$

We called it "simple" because we are using only one predictor variable *X*. The ***slope*** β_1 or ***population regression slope*** tells us how much the value of the response *Y* changes when the value *X* changes one unit. If the population regression slope is zero, then the variable *X* does not have any linear relation with the response variable *Y*. The ***intercept*** β_0 or ***population regression intercept*** is the value of the response *Y* when the value of *X* is zero. To model the departure from a perfect line, we introduce

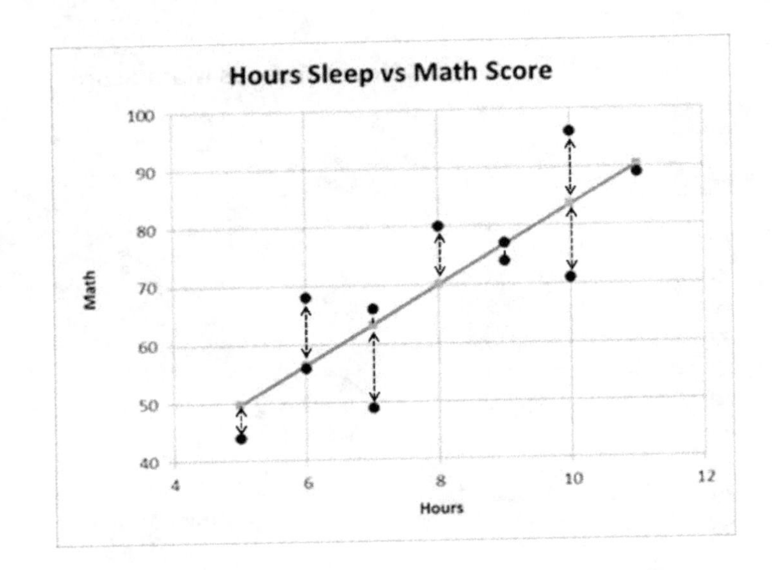

Figure 18.3.

the ***error of prediction*** ε_i. Individual observations have different errors of prediction and the variability of these errors is the ***variance of error term*** σ_ε^2.

Of course, we rarely have access to the whole population data, thus we compute the regression equation using a sample with N observations. The ***sample regression line equation*** is given by:

$$\hat{Y}_i = b_0 + b_1 X_i \qquad VAR(e_i) = MSE$$

where b_0 is the ***sample regression intercept***, b_1 is the ***sample regression slope***, and MSE is the ***sample mean square error*** or variability of the prediction errors, e_i, that in the sample are referred to as ***residuals***. The ***predicted value*** $\hat{Y}$ is the value of Y that the equation predicts by using the regression equation. Of course, this predicted value is not necessarily the actual value of Y that corresponds to an observed value of X, and therefore the residual is defined as $e_i = Y_i - \hat{Y}_i$, i.e., the difference between the observed and the predicted value of the response. To obtain the equations for the slope and the intercept, we put two restrictions: (1) the point $(\bar{X}, \bar{Y})$ should be on the line and (2), as we mentioned before, the sum of squared residuals should be as small as possible.

The sample regression slope is computed as the ratio of the ***sample covariance***, S_{XY}, between X and Y over the sample variance of X, S_X^2. If the covariance is close to zero, there is no linear relation between the two variables. An alternative equation for the regression slope uses the ***sum of cross -products, SS_{XY}***, over the sum of squares of the predictor variables X, or ***sum of squares of X, SS_X***. Once we compute the regression slope, the regression intercept is the difference between the average of Y and the product of the slope and the average of X. Thus, the formulas for the slope and the intercept of the simple least square regression are:

$$b_1 = \frac{S_{XY}}{S_X^2} \quad or \quad b_1 = \frac{SS_{XY}}{SS_X} \quad and \quad b_0 = \bar{Y} - b_1\bar{X}$$

where $SS_X = \sum_{i=1}^{n}(X_i - \bar{X})^2$

Of course, the sample slope and intercept are statistics, i.e., their values may change from sample to sample. We apply the same inferential statistics procedures to set confidence intervals and hypothesis testing for the slope and the intercept. The usual ***test for the regression slope*** assesses if the sample regression slope comes from a population with slope zero, i.e., we test the hypotheses $H_0 : \beta_1 = 0$ versus $H_1 : \beta_1 \neq 0$. Under the usual assumption of normality, the test-statistic, t_{obt}, for testing the regression slope is:

$$t_{obt} = \frac{b_1}{S_{b_1}}$$

where b_1 is the sample regression slope, and S_{b_1} is the standard error for the regression slope, for which the formula is:

$$S_{b_1} = \sqrt{\frac{SSE/(n-2)}{SS_X}} = \sqrt{\frac{MSE}{SS_X}}$$

where $SSE = \sum_{i=1}^{n}(Y_i - \hat{Y}_i)^2$, the ***sum of squares error***, is the variability due to the difference between the actually observed values of the response Y and the predicted values $\hat{Y}$ when using the regression equation. The SSE can also be computed by using the sum of squares of X and Y, and the sum of cross products:

$$SSE = SS_Y - \frac{SS_{XY}^2}{SS_X}$$

MSE, the ***mean square error***, is the sum of squares error divided by $n - 2$, the degrees of freedom of the regression. The sampling distribution for the test-statistic is the t-distribution with $N - 2$ degrees of freedom.

The 100(1 – α)% confidence intervals for the population regression slope becomes:

$$b_1 \pm t_{\alpha/2;N-2} S_{b_1}$$

Similarly, we can obtain standard error for the regression intercept and obtain a test and confidence interval for the intercept. You can find the formulas in Section 18.3 of this chapter, where we describe regression computation using basic Excel functions.

Finally, how well the regression line describes the data, or how large the magnitude of the relation between X and Y is, i.e., what the effects size is, is described by the ***coefficient of determination, or r-squared***. The value of r-squared expresses the *proportion* of the total variability in the response Y that X helps to reduce or account for. Because r-squared is a proportion, we can also express it as a *percentage* of the total variability of Y that X accounts for. The larger the value of r-squared, the better. ***The r² is actually the square of the correlation r between X and Y.***

The value of r-squared is computed as follows:

$$r^2 = 1 - \frac{SSE}{SS_Y}$$

where $SS_Y = \sum_{i=1}^{n} (Y_i - \bar{Y})^2$ is the sum of squares of the response Y.

As you can see, the computations for linear regression are more involved than any other previous techniques we have seen. Fortunately, Excel can display the regression line equation in scatterplots produced by the Chart menu. Also, we will extensively rely on the Data Analysis ToolPak to obtain linear regression tests and confidence intervals. Nevertheless, we will also present one example of how to compute a simple regression using basic Excel functions.

18.1 OBJECTIVES

1. To use the Chart menu to obtain scatterplots and plots of simple regression lines.
2. To compute a simple regression line using basic Excel functions.
3. To compute a simple regression line using the Data Analysis ToolPak.

SCATTERPLOT AND SIMPLE REGRESSION EQUATION

18.2

In a job absenteeism study, researchers investigated the relationship between the number of cigarettes smoked by employees and their job absenteeism. The number of cigarettes smoked daily and the number of days absent from work during one year due to illness were obtained for a sample of 12 individuals employed in a manufacturing plant. The data is in the first worksheet of the Excel file for this chapter.

The first step of a regression analysis is to obtain the scatterplot for the data (see Figure 18.4).

- Highlight the columns B and C containing the data, including the headers (range B1:C13).
- Click the Insert tab in the Excel top menu. In the Charts click on the Scatter icon and select the Scatter option. This will create the basic scatterplot.
- In the Excel top menu, you will see a Chart Tools tab with two options: Design and Format. Click on the Design tab and the "Quick Layout" icon. Select the first layout option. This will allow you to enter titles for the x (cigarettes smoked) and y (days absent) axes. We can erase the overall title of the chart (the *Y*-Absent label) and the legend on the right-hand side of the plot.

To overlay the regression line with the regression equation and the r-squared value:

- Click on the scatterplot and on the Plus icon that will appear on the right-hand side of the plot.
- Check the option "Trendline" (the regression line will appear on the scatterplots).
- Click on the right arrow next to the "Trendline" option, and select "more options." This will open the Format Trendline menu.
- Check the boxes "Display Equation on chart" and "Display R-squared value on chart."

The regression equation has a slope of 0.1636 and an intercept of 4.7006. The value of the r-squared is 0.456. Thus, the slope shows that for each additional cigarette per day, we estimate an increase of 0.1636 absent working days. In addition, by knowing the number of cigarettes smoked per day by an employee, we can account for 45.6% of the variability in job absenteeism. This is a seemingly good fit for the model; however, we still have to test if this slope is significantly different from zero.

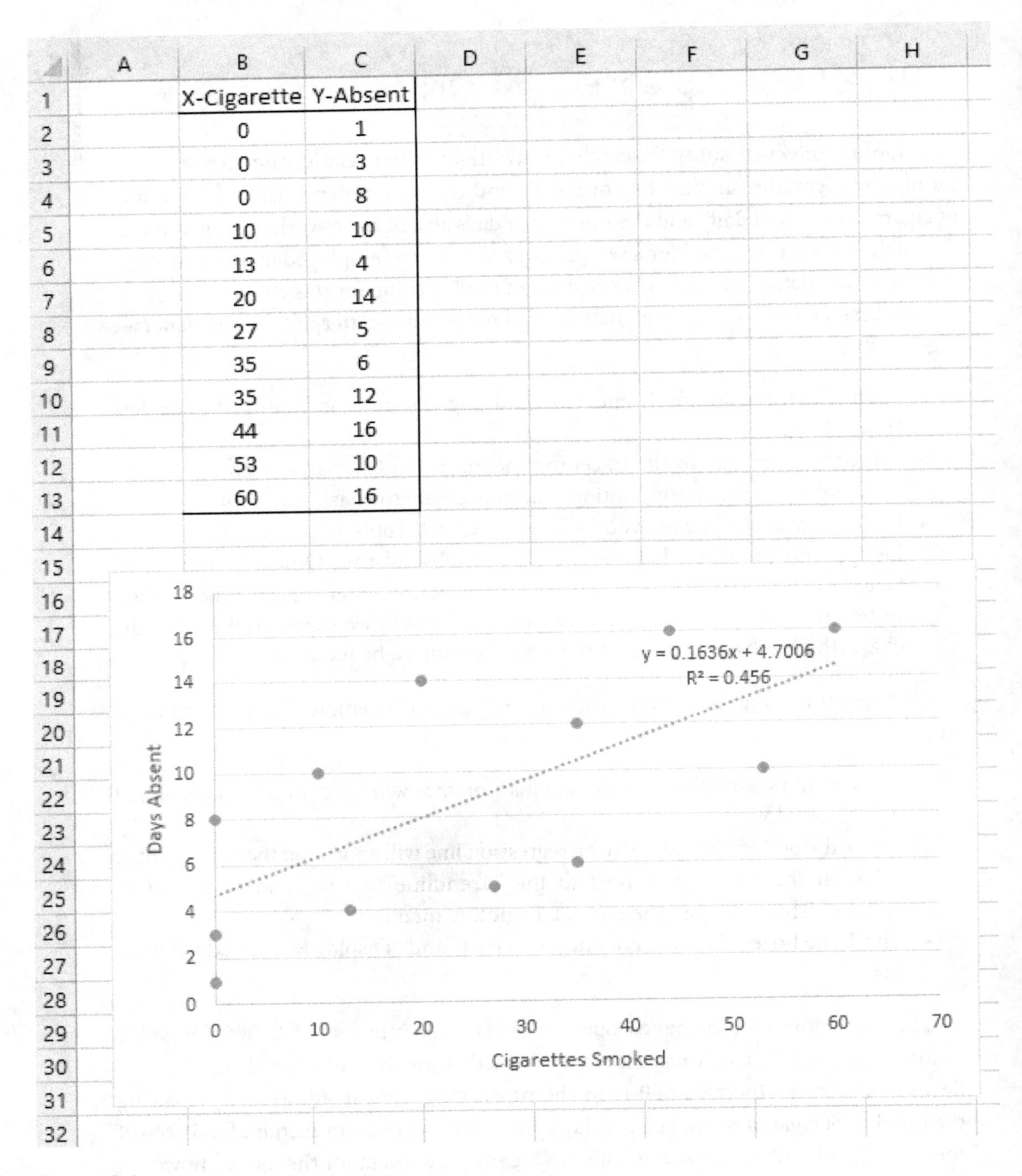

X-Cigarette	Y-Absent
0	1
0	3
0	8
10	10
13	4
20	14
27	5
35	6
35	12
44	16
53	10
60	16

Figure 18.4.

SIMPLE LINEAR REGRESSION USING EXCEL FUNCTIONS

18.3

In the second worksheet of the Excel file for this chapter, we compute the regression equation in such a way as to mimic computations by hand (see Figure 18.5).

	A	B	C	D	E	F	G
1		X-Cigarette	Y-Absent				
2		0	1				
3		0	3				
4		0	8				
5		10	10				
6		13	4				
7		20	14				
8		27	5				
9		35	6				
10		35	12				
11		44	16				
12		53	10				
13		60	16				
14	Mean	24.75	8.75				
15	SD	20.98105206	5.083395429				
16	N	12	12				
17	SS	4842.25	284.25				
18	Cov			72.02272727			
19	SSXY			792.25			
20							
21							
22	SSE	154.6284269					
23	MSE	15.46284269					
24							
25		Coefficient	SE	tobt	p	Low 95%	Up 95%
26	bo	4.700604058	1.801298672	2.60956394	0.026060286	0.6870605	8.71414761
27	b1	0.163611957	0.056509448	2.895302699	0.015961514	0.03770106	0.28952285
28							
29	r2	0.45601257					
30							

Figure 18.5.

First, we compute the mean, standard deviation, sample sizes, and the sum of squares for the variables X and Y. The sum of squares for X can be obtained by $SS_X = (N-1)S_X^2$ and for Y in a similar formula.

- In cell B14, enter =AVERAGE(B2:B13), and drag also the equation to C14.
- In cell B15, enter =STDEV.S(B2:B13), and drag also the equation to C15.
- In cell B16, enter =COUNT(B2:B13), and drag also the equation to C16.
- To compute SS_X, in cell B17, enter =(B16-1)*B15^2. Drag the equation to C17 to obtain SS_Y.

The covariance between variables X and Y, S_{XY}, can be obtained by the Excel function COVARIANCE.S. The sum of cross products, SS_{XY}, is easily obtained as $SS_{XY} = (N-1)S_{XY}$.

- For covariance, enter in cell D18, the function =COVARIANCE.S(B2:B13, C2:C13). Notice that the input consists of two range of values, the one for X and the one for Y.
- For sum of cross products, enter in cell D19, the equation =(B16-1)*D18.

The SSE is computed as $SSE = SS_Y - SS_{XY}^2/SS_X$ and the MSE as $MSE = SSE/(n-2)$:

- Enter in B22, the equation =C17 – D19^2/B17.
- Enter in B23, the equation =B22/(B16-2).

The remaining values are computed using the sum of squares, sum of cross products, and means. These values are put together in a table. We start with the formulas for slope b_1:

$$b_1 = SS_{XY}/SS_X \quad S_{b_1} = \sqrt{MSE/SS_X} \qquad t_{b_1} = b_1/S_{b_1} \quad p\left(t_{n-2} > |t_{b_1}|\right) \quad b_1 \pm t_{\alpha/2;n-2} S_{b_1}$$

- To compute slope, b_1, in cell B27, enter =D19/B17.
- To compute the standard error for the slope, S_{b1}, in cell C27, enter =SQRT(B23/B17).
- To compute the t-obtain for the slope, t_{b1}, in cell D27, enter =B27/C27.
- To obtain the p-value, in cell E27, enter =T.DIST.2T(D27, B16-2).
- To obtain the lower limit for the 95% CI, enter in cell F27 =B27-T.INV.2T(.05, B16-2)*C27.
- To obtain the upper limit for the 95% CI, enter in cell G27 = B27+T.INV.2T(.05, B16-2)*C27.

The computations for the intercept $b0$ of the regression use the formulas:

$$b_0 = \bar{Y} - b_1\bar{X} \quad S_{b_0} = \sqrt{MSE(1/n + \bar{X}^2/SS_X)} \quad t_{b_0} = b_0/S_{b_0} \quad p\left(t_{n-2} > |t_{b_0}|\right) \quad b_0 \pm t_{\alpha/2;n-2} S_{b_0}$$

- To compute intercept b_0, in cell B26, enter =C14-B27*B14.
- To compute the standard error for the intercept, S_{b0}, in cell C26, enter =SQRT(B23*(1/B16-B14^2/B17)).
- To compute the t_{obt} for intercept t_{b0}, in cell D26, enter =B26/C26.
- To obtain the p-value, in cell E26, enter =T.DIST.2T(D26, B16-2).
- To obtain the lower limit for the 95% CI, enter in cell F26 =B26 - T.INV.2T(.05, B16-2)*C26.
- To obtain the upper limit for the 95% CI, enter in cell G26 =B26 + T.INV.2T(.05, B16-2)*C26.

Finally, to compute r-squared, we use the formula $r^2 = 1 - SSE/SS_Y$

- In cell B29, enter the equation = 1-B22/C17.

The regression intercept and slope, and the r-square, are the same as the ones reported in the scatterplot. However, now we compute the test for the slope (as well as for the intercept). For an alpha of .05, we can reject the null hypothesis that this sample comes from a population with a regression slope zero, (p = .01596151). Thus, there is a significant relation between the number of cigarettes smoked per day and job absenteeism.

18.4 LINEAR REGRESSION USING THE DATA ANALYSIS TOOLPAK

By far, a more practical way to obtain a simple linear regression in Excel is by using the Data Analysis ToolPak. In the third worksheet in the Excel file for this chapter, we have again the data set and the basic regression output from the Data Analysis ToolPak (see Figure 18.6).

- Click on the Data tab in the Excel top menu and then on the Data Analysis icon.
- Select the option "Regression" and then OK.
- Put as input Y range the values in column C, including the header (range C1:C13).

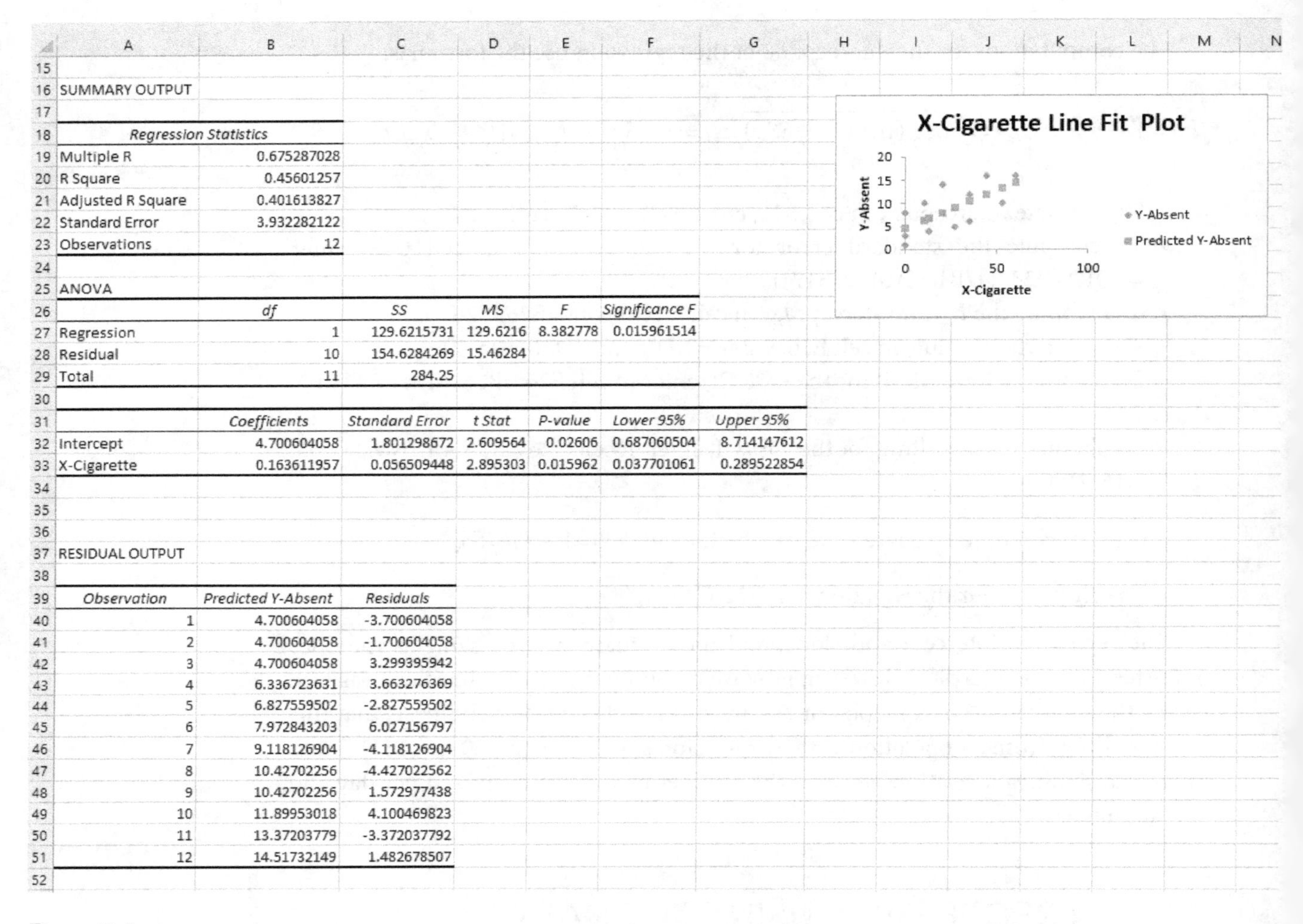

SUMMARY OUTPUT

Regression Statistics	
Multiple R	0.675287028
R Square	0.45601257
Adjusted R Square	0.401613827
Standard Error	3.932282122
Observations	12

ANOVA

	df	SS	MS	F	Significance F
Regression	1	129.6215731	129.6216	8.382778	0.015961514
Residual	10	154.6284269	15.46284		
Total	11	284.25			

	Coefficients	Standard Error	t Stat	P-value	Lower 95%	Upper 95%
Intercept	4.700604058	1.801298672	2.609564	0.02606	0.687060504	8.714147612
X-Cigarette	0.163611957	0.056509448	2.895303	0.015962	0.037701061	0.289522854

RESIDUAL OUTPUT

Observation	Predicted Y-Absent	Residuals
1	4.700604058	-3.700604058
2	4.700604058	-1.700604058
3	4.700604058	3.299395942
4	6.336723631	3.663276369
5	6.827559502	-2.827559502
6	7.972843203	6.027156797
7	9.118126904	-4.118126904
8	10.42702256	-4.427022562
9	10.42702256	1.572977438
10	11.89953018	4.100469823
11	13.37203779	-3.372037792
12	14.51732149	1.482678507

Figure 18.6.

- Put as input X range the values in column B, including the header (range B1:B13).
- Check the "Labels" box.
- Check the Output Range option, and select as first cell for the outcome A16.
- Check the option "Line Fit Plots." Then press OK.

Excel will produce four tables. The first table contains, among other values, the "R Square" for the regression. The value of 0.45601257 is the same one that we found before using direct computations.

The second table is an Analysis of Variance table. This table is used to test the overall hypothesis that *all* regression slopes are zero when we run a regression with more than one predictor variable, i.e., when we run multiple linear regression. In the case of a simple linear regression, we can see that the entry for "Residual" in the SS column is the value of 15.46284 for SSE that we found before when using direct computations. In addition, the "Residual" degrees of freedom, 10, is the degrees of freedom that we use for the t-tests for intercept and slope.

The third table is the same as we computed before using direct computations. The "X-Cigarette" row contains the information for the regression slope coefficient, the standard error, the t_{obt}, the p-value, and the 95% CI for the slope.

The "Line Fit Plots" option produced the fourth table and the scatterplot with the predicted values for the regression. The fourth table contains the predicted values, $\hat{Y}$, and the residuals e (the difference between the observed values of Y and the predicted values) for each of the 12 observations in the data. The plot is a rough representation of the regression line; however, we can edit it at will.

WORKED EXAMPLE

18.5

In the fourth worksheet in the Excel file for this chapter, we have the ages of 169 British married couples that were collected in 1980. There are two columns in the data set: "Husband Age" and "Wife Age." We would like to know the relationship between husband and wife ages. In concrete, we would like to predict the age of the wife (as Y) knowing the age of the husband (as X).

1. We obtain and report the scatterplot of wife age (as variable Y) over husband age (as variable X) and overlay the regression line with the equation and r-squared. (a) Do you think that there is a linear trend in the scatterplot, positive or negative? (b) Interpret the regression slope. (c) Report the r-square. Based on this value, is there a good fit between the ages of spouses? (see Figure 18.7).
    - Highlight columns A and B containing the data, including the headers (range A1:B170).
    - Click the Insert tab in the Excel top menu. In the Charts tab click on the Scatter icon and select the Scatter option. This will create the basic scatterplot.
    - In the Excel top menu, select the "Design" option in the Chart Tools tab and then click the "Quick Layout" icon. Select the first layout option. Enter "Husband Age" as x-axis title, and "Wife Age" as y-axis title. Erase the overall title of the chart.

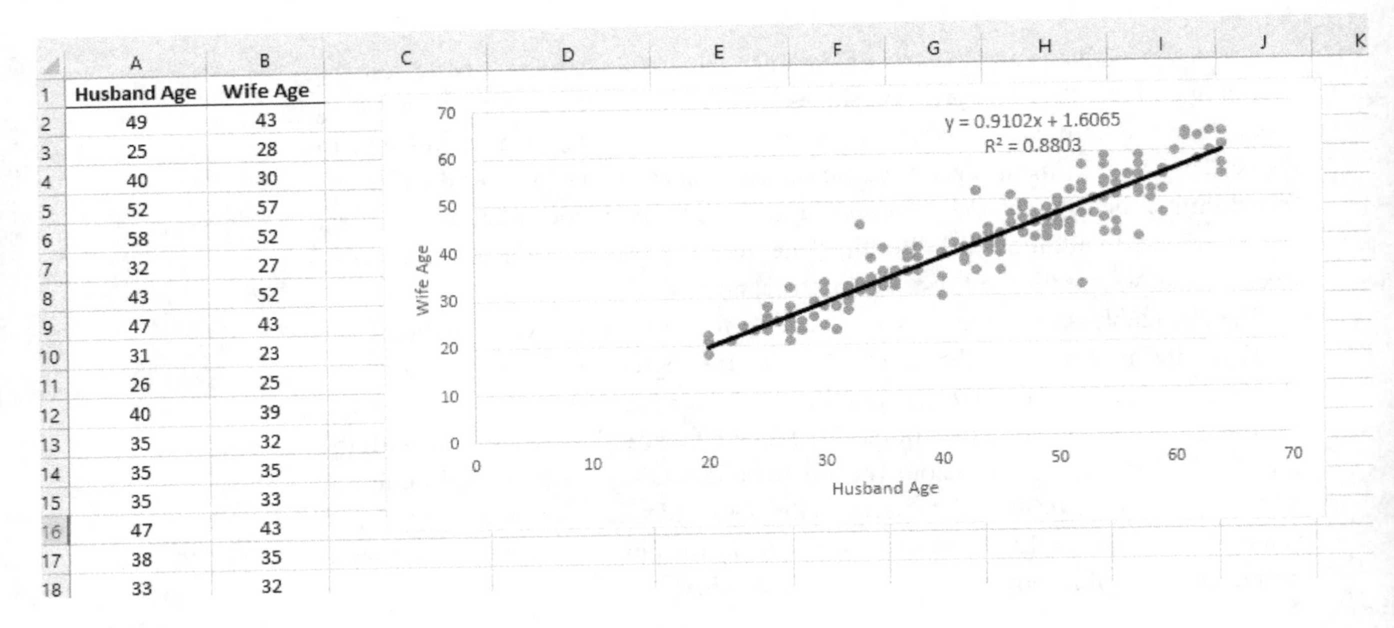

Husband Age	Wife Age
49	43
25	28
40	30
52	57
58	52
32	27
43	52
47	43
31	23
26	25
40	39
35	32
35	35
35	33
47	43
38	35
33	32

Figure 18.7.

- Click on the scatterplot and on the Plus icon that will appear on the right hand side of the plot.
- Check the option "Trendline" (the regression line will appear on the scatter-plots).
- Click on the right arrow next to the "Trendline" option, and select "more options." This will open the Format Trendline menu.
- Check the boxes "Display Equation on chart" and "Display R-squared values on chart."
- To make the trendline more clear, click twice on the line. The "Format Trend-line" menu will appear. Click on the paint bucket icon and change the color of the line to black and its shape to solid line.
- The equation and R-squared values appear inside a text box. To make it more readable, click on the box and increase the font size. Also, move the box outside the cloud of dots.

a) Yes, there is a clear positive linear trend in the data.

b) The regression equation states that for each additional year of husband age, the age of the wife increases by 0.9102 years.

c) Based on the *r*-square of 0.8803, there is a good fit between the ages of spouses.

	A	B	C	D	E	F	G	H	I
22	59	55							
23	26	25	SUMMARY OUTPUT						
24	50	45							
25	49	44	*Regression Statistics*						
26	42	40	Multiple R	0.938238173					
27	33	31	R Square	0.88029087					
28	27	25	Adjusted R Squar	0.879574049					
29	57	51	Standard Error	3.959062034					
30	34	31	Observations	169					
31	28	25							
32	37	35	ANOVA						
33	56	55		*df*	*SS*	*MS*	*F*	*Significance F*	
34	27	23	Regression	1	19248.6381	19248.64	1228.048	6.93262E-79	
35	36	35	Residual	167	2617.586755	15.67417			
36	31	28	Total	168	21866.22485				
37	57	52							
38	55	53		*Coefficients*	*Standard Error*	*t Stat*	*P-value*	*Lower 95%*	*Upper 95%*
39	47	43	Intercept	1.606499089	1.153900305	1.392234	0.165703	-0.67161277	3.8846109
40	64	61	Husband Age	0.910176975	0.025972763	35.04352	6.93E-79	0.858899704	0.9614542
41	31	23							

Figure 18.8.

2. Use the Data ToolPak to obtain the regression analysis output for wife age (as variable Y) over husband age (as variable X). (a) Report and interpret the value of the regression slope. (b) Report the t_{obt} value for the test of the slope, the corresponding degrees of freedom, and the p-value (round up values to four decimals). (c) Using an alpha of .05, is the regression slope significantly different from zero? (see Figure 18.8).
   - Click on the Data tab in the Excel top menu and then on the Data Analysis icon.
   - Select the option "Regression" and then OK.
   - Put as input Y range the values in column B, including the header (range B1:C170).
   - Put as input X range the values in column A, including the header (range A1:A170).
   - Check the "Labels" box.

- Check the Output Range option, and select as first cell for the outcome C23. Then press OK.
  a) The regression slope is 0.9102. Thus, for each additional year in the age of a husband, we predict that the age of the wife will increase by 0.9102.
  b) The t_{obt} is 35.0435, with 167 df (the degrees of freedom of the residual in the ANOVA table) and p-value $< .0001$.
  c) We reject the null hypothesis that the slope comes from a distribution with mean zero. The regression slope is significantly different from zero.

The result of the analyses is in worksheet 5 of the lab Excel file.

CPSIA information can be obtained
at www.ICGtesting.com
Printed in the USA
LVHW060100091222
734861LV00003B/42

9 781516 599080